BREAKWATERS

Design & Construction

Proceedings of the conference organized by the Institution of Civil Engineers and held in London on 4–6 May 1983

Thomas Telford Limited, London

Published for the Institution of Civil Engineers by Thomas Telford Ltd, PO Box 101, 26–34 Old Street, London EC1P 1JH

First published 1984

Conference organizing committee: P. Lacey, W. A. Price, R. G. Tickell and I. W. Stickland

British Library Cataloguing in Publication Data
Breakwaters
  1. Breakwaters
  I. Institution of Civil Engineers
627'.24   TC333

ISBN 0 7277 0190 8

Printed by The Thetford Press Limited, Thetford, Norfolk

Contents

THEME PAPER: State of the art. I. W. STICKLAND   1

SESSION 1: WAVE CLIMATE

1. Synthesis of design climate. J. C. BATTJES   9

2. Specification of construction climate. M. A. MESTA   19

Discussion on Papers 1 and 2   25

SESSION 2: MATERIALS FOR CONSTRUCTION

3. Durability of rock in breakwaters. A. B. POOLE, P. G. FOOKES, T. E. DIBB and D. W. HUGHES   31

4. Concrete and other manufactured materials. A. R. CUSENS   43

Discussion on Papers 3 and 4   49

THEME PAPER: The design process. J. E. CLIFFORD   53

SESSION 3: FOUNDATION PROBLEMS AND MODELLING

5. Foundation problems. W. R. THORPE   65

6. Hydraulic modelling of rubble-mound breakwaters. M. W. OWEN and N. W. H. ALLSOP   71

Discussion on Papers 5 and 6   79

SESSION 4: ARMOUR STABILITY

7. Field scale studies of riprap. P. ACKERS and J. D. PITT   91

8. The core and underlayers of a rubble-mound structure. T. S. HEDGES   99

9. The design of armour systems for the protection of rubble-mound breakwaters. W. F. BAIRD and K. R. HALL   107

Discussion on Papers 7–9   121

SESSION 5: RISK ANALYSIS

10. Risk analysis in breakwater design. A. MOL, H. LIGTERINGEN and A. PAAPE   133

Discussion on Paper 10   139

SESSION 6: SPECIFICATION AND CONSTRUCTION

11. Construction methods and planning. J. F. MAQUET   143

12. Rubble breakwaters—specifications. P. G. R. BARLOW and M. G. BRIGGS   151

Discussion on Papers 11 and 12   159

SESSION 7: MAINTENANCE, REPAIR AND MONITORING EQUIPMENT

13. Monitoring of rubble-mound structures. O. T. MAGOON, V. CALVARESE
    and D. CLARKE   167

Discussion on Paper 13   173

THEME PAPER: The way ahead. H. F. BURCHARTH   177

Theme paper: State of the art

I. W. STICKLAND, Posford, Pavry & Partners

An overall view is taken of changes in the state of the art of breakwater design and construction over the last 40 years. Particular reference is made to the historical development of concrete armour units and questions raised as to their suitability in the light of recent failures. The absence of adequate design guidelines is highlighted and the need for more research emphasised.

Any discussion about the 'State of the Art' of the design and construction of breakwaters in 1983 must inevitably be done in the knowledge that the last 4-5 years have been dominated by reports of damage or failure of rubble mound breakwaters in depths of water greater than 15 metres. I refer, of course, to Sines, Arzew, San Ciprian and Bilbao. This, however, is a far too one-sided picture and it is wholly wrong to believe that these problems are confined to structures in deeper water.

In a recent world wide survey of existing breakwaters, initiated originally by the United Kingdom and continued later under the banner of P.I.A.N.C., detailed reports were received on 148 structures; all of the structures were either of the composite or rubble mound type. Not all these reports indicated the depth of water but of the 136 that did, 101 were in depths of less than 15 metres and 35 in depths equal to or greater than 15 metres.

Reports of damage to these structures varied from slight to 100% and occurred to 30 break-waters in depths less than 15 metres and 10 in depths equal to or greater than 15 m. It is also worth noting that a great many of the structures which suffered damage have been in existence for less than 15 years.

Whilst the P.I.A.N.C. review does not pretend to be all embracing and clearly could not hope to cover all breakwaters, there is, I suggest, sufficient evidence to give cause for concern.

This concern was recognised by P.I.A.N.C. by the formation in 1981 of an International Working Group who have specific terms of reference which required them to review all aspects of the design, construction and maintenance of rubble mound structures. The report mentioned previously forms part of that review. The Working Group set about this task by considering the various problems under the following 7 headings.

1. Wave Climate
2. Wave/Structure Interaction
3. Model Studies
4. Geotechnical Aspects
5. Strength of Materials
6. Construction Procedures
7. Design Philosophy

The work is now virtually complete and a final draft of the report to P.I.A.N.C. will be discussed and agreed at a meeting of the Working Group to be held immediately after this conference.

The report in itself is essentially another statement of the 'State of the Art' and as a member of that Working Group I must inevitably lean heavily on the contents as a foundation for this paper. This, of course, must not be taken to imply that anything I say or any opinions I express are necessarily shared by my colleagues in the Working Group. It has, after all, been said, and on many occasions, that 'two harbour engineers can hold diametrically opposite views and both be right', nor indeed can I hope to cover, within the limitations of one paper, the full range of problems facing us.

In a conference such as this, all of the papers represent statements on the 'State of the Art', and having actually read the papers, I believe the authors have done it very well. I must also say at the outset that to attempt to cover the whole range from the design, construction and maintenance on all types of breakwater within one paper, is clearly impossible. I propose therefore to take a broad view of the situation and highlight what I believe to be some critical areas and to look briefly at the history of the development of the Art of Breakwater Design, particularly over the last 40 years.

Right at the outset I think we must all be aware that the structures we are dealing with are in many ways far removed from other civil engineering structures. This has been said before but I believe it is worth repeating.

Structural frames, be they buildings, or bridges, can be designed and built with a degree of confidence not generally available to the harbour engineer.

The various loading cases are clearly defined, the structural shapes are amenable to rigorous mathematical anlysis, and the physical characteristics of the materials employed are well researched and understood. Almost the entire process is embodied in various Codes of Practice. Furthermore, the degree of control, precision and inspection which can be employed during the construction on land greatly enhances the chances of

Breakwaters—design and construction. Thomas Telford Ltd, London, 1984

1

long term structural integrity. The process is, in short, predictable.

By comparison, a breakwater generally consists of a large percentage of randomly placed elements, frequently lacking in cohesion or homogeneity, subject to random wave forces which, in themselves, result in physical processes that are not fully understood. If this were not enough by itself, the designer must also recognise that the shape of the structure and the disposition of the various elements comprising that structure may well change with time and in so doing, depart significantly from the original design assumptions. The problem is further compounded in that the control, precision and quality inspection which can be undertaken on a structure, 9/10th of which is underwater, presents its own difficulties and uncertainties.

Given, therefore, this general air of uncertainty, it is clear that both the designer and the contractor are having to accept a level of risk which, in other branches of civil engineering, would be considered as wholly unacceptable.

The problems start with the determination of the design loading, in this case, wave conditions, and more importantly the wave conditions which will be experienced throughout the life of the structure. In many cases, breakwaters will be the first stage of a very long term development where they form in effect a shell within which various types of berths and facilities . such as dry docks etc. can ultimately be constructed.

Far too often a designer when faced with the problem of determining wave conditions in such a situation finds that there is little or no data directly applicable to his site. Almost inevitably there will be no data based on actual measurements of waves.

This is perhaps not too surprising when one realises that there were very few wave recorders available on a commercial basis until the 1960s and reliability required another 10 years of intensive development. Even if he installs a wave recorder soon after receiving instructions to proceed the overall time scale for the design phase is unlikely to result in much more than one year's worth of data. Furthermore, this data will not provide him with any information on wave direction. He therefore has to resort to an examination and analysis of other sources of data such as visual observations of ships in passage, wind data from shore stations based on synoptic weather charts and in certain cases mathematical modelling. On the basis of this data, he must attempt to define the extreme conditions likely to be experienced, the 100 year storm for instance. Regrettably, different extreme value analysis methods yield different answers and there is a need for more research into this aspect. Nor will he necessarily be able to establish by any of the methods above other important wave parameters such as wave grouping, sequences of large waves, long period waves, the possibility of exceptional large 'freak' waves, shallow water effects and many others. Equally, he must, if he wants to optimise his design, seek to establish the frequency of occurrence of all these various phenomena accepting in the process that many of the phenomena are not independent variables.

In short, even if the designer uses all the available sources and methods presently to hand, the confidence level he can have in the determination of his design loading falls far short of what would be expected by his colleagues in structural engineering.

Nevertheless, whoever it is who now finds himself in this position, he has to begin to decide which type of breakwater will best meet his requirements. Basically, he has three options.

1. A vertical wall structure
2. A composite structure consisting of a large concrete superstructure sitting on a mound of rubble
3. A complete rubble mound rising above water level

Many factors will affect his decision and I do not propose to go into them in any detail here as this will be dealt with much more adequately, by John Clifford in his keynote address on the 'Design Process'. I will instead confine myself to some more general considerations starting with the gravity wall structure and in this instance, I am concerned with structure sitting on a minimal thickness of rubble on the sea bed.

Considerable development on the technology of vertical faced gravity wall structures has taken place over the last two decades principally in Japan and Denmark. In the case of Japan the lack of sufficient good rock has forced them to adopt this approach and they probably have more vertical faced breakwaters than any other country in the world. It should not however be forgotten that throughout the 19th century this type of structure was almost exclusively used throughout Europe from fishing harbours to major ports. Bearing in mind the very low level of technology available to the harbour engineer of that time, it is interesting to observe how many of these structures have survived, in a good state of preservation, to this day. The majority of these older breakwaters were constructed using either masonry or concrete blocks laid horizontally or in sloping slice work patterns on good solid foundations, and sometimes in depths of water exceeding 20 metres.

The advent of reinforced concrete led eventually to the replacement of blockwork with concrete caissons cast on shore floated into position and lowered on to a carefully prepared bed. This approach has now reached a stage when vertical breakwaters can be at least as competitive as rubble mound designs, particularly if good quality rock is not available near the site. There is a further very important advantage and that is that the ability to construct a major portion of the structure on dry land enables a much greater degree of quality control and inspection to be achieved than is possible in an underwater construction situation. In addition, provided the wave forces on the structure are rigorously investigated, the subsequent stress analysis can proceed along conventional lines with a resulting increase in the level of confidence.

There are, of course, some major disadvantages, not the least of which is the question of wave reflections and much research has gone into devising means of reducing this. The methods employed have ranged from hollow caissons having outer walls with perforations or slits to complex shaped individual blocks with internal passage-ways to create turbulence and thus dissipate energy. The efficiency of these devices tends to be uniquely sensitive to wave period with a sometimes very rapid rise in the reflection coefficient either side of the discreet design conditions. Equally, the loss of overall weight arising from the hollow chambers can present stability problems and has to be compensated for in some fashion. Other solutions to this problem of reflection involve the placing, on the sea-ward face, of large mounds of concrete blocks or armour units with the attendant underlayers and core resembling in total half a rubble mound breakwater. The logic of this approach is difficult to appreciate as the requirements of the mound dictate a supply of rock equivalent to that of a conventional rubble mound form of construction which the caisson structure was, of course, intended to replace. Presumably, due to wave action, there is an argument that the reduc-tion of the horizontal forces on the vertical wall resulting from the presence of the mound enables the overall dimension of the caisson to be reduced.

Yet another method of reducing reflection and in the process reducing the horizontal forces on the breakwater is the concept developed in Denmark of a sloping face on the seaward side above the water level at an angle to the horizon-tal of some 30°. The gains obtained in the reduction of reflections and wave forces have, however, to be offset by the resulting increase in rates of overtopping and the distinct possi-bility of generating unwanted wave activity within the berthing area.

The type of construction adopted in the 19th century of massive solid concrete blockwork structures founded on rock were carried out in the absence of hydraulic model tests. Such a course of action cannot be contemplated in the more recent designs particularly as many of the breakwaters have to be founded on softer and weaker material. Modern day designs are heavily dependent on very extensive and sophis-ticated model studies which, if they are not to be dangerously misleading, require to be based on detailed wave recordings over as long a period as possible. One only has to look at the sort of problem which can arise. Shock pressures are extremely sensitive to the pro-file of the wave. Neither regular nor irregular round crested waves produce shock forces and it is therefore insufficient to produce electroni-cally in a flume a correct round crested wave spectrum without introducing wind which, in turn, may have to be exaggerated in order to produce sharply crested profiles.

The model tests must also take account of:

(a) The pressures in an air pocket trapped between the structure and the concave wave front

(b) The pressure in bubbles entrained in the breaking wave as well as the concentration of entrained air

(c) The pressure in the air cushion that is being expelled when the wave front collides with the structure

(d) Interaction between wave forces and forces induced in the underlying soil due to the rocking motion of the structure on its foundations under wave action

Assuming that these factors can be accommodated, the frequency response of the measuring instru-ments must be very high to cope with very rapid rise in shock pressures.

As these shock pressures are functions of the ever varying combinations of wave shapes, it follows that the results must be statistically analysed to give wave force distributions for various water levels, significant wave heights to the joint distribution of combinations of water levels and storm value of H_s from the pro-totype wave recording.

One final factor needs to be mentioned before passing on to composite type breakwaters and that is the question of scour at the sea bed arising from the very complex wave action associated with vertical structures. It is clear-ly a very critical element in the design and yet, as far as I am aware, there are no published guidelines to help the practising engineer deter-mine the size, grading, and extent of this scour apron. Again he must resort to model studies.

Much that I have said on full depth gravity wall breakwaters applies to composite structures with the added complication of the stability of the frequently very large rubble mound on which the superstructure sits. Again, the majority of experience with this type of structure in recent times has been in Japan, although many composite breakwaters were built around the Mediterranean at the beginning of this century, sometimes in depths of water as great as 30 to 40 metres. The superstructures of these earlier breakwaters were constructed using very large concrete blocks, weighing as much as 200 to 400 tons each. Failures did occur in some instances due to the blocks being bodily removed by wave action or a failure of the toe mound. The design of the latter is critical as it can lead to plunging breakers hitting the superstructure unlike many full depth gravity walls which do not experience this phenomenon. Once again there are no published guide lines to help the engineer em-barking on such a design. He will be forced to undertake a detailed literature search to find the many papers covering the extensive and excellent research carried out mainly in Japan. Indeed there is a very urgent need for all this information to be collected evaluated and brought together in one volume. Until this is done he will experience great difficulty in developing even his first tentative cross section which will form the basis of his model test programme. It is absolutely certain that he cannot proceed without hydraulic model studies which will be no less intensive than in the case of the gravi-ty wall. They will have the additional compli-cation of the stability of the rubble mound and in particular the very complex conditions at the base of the superstructure where the mound must withstand the very large downward forces due to wave action.

Unless the designer can be certain of obtain-ing adequate and sufficient wave records from the site in question extending over a period of years, the risk level associated with this type of breakwater is very high indeed.

SUMMARY

We now come to the type of breakwater which is the centre of more argument, discussion, diversity of opinion and disagreement than almost any other maritime structure. I refer of course to the rubble mound breakwater.

What then is a rubble mound breakwater? It is in fact all things to all men.

To the chairman of a Port Authority it is a large heap of rock dumped, at great expense, in the sea to protect an area of water, which he, the Client, wants to use at all times and in all weather because he has sold his harbour to shipping companies as providing just that. He sometimes fails to understand why engineers and scientists contrive to make the design sound so difficult and sophisticated because even he, as a layman, knows that breakwaters have been around for hundreds and hundreds of years. Why therefore, he asks himself, do these damned things still fall apart and who is this chap Hudson who keeps on cropping up in every conversation?

To a Mariner it is a navigational hazard which on occasions or when designed by an inexperienced engineer, somehow seems to make the seas greater inside the harbour than out. He has been told that it is rather like an iceberg with 9/10th of it under water - somewhere - so he gives it a wide berth and ends up driving his ship on to the Lee breakwater which, of course, was put there for just that purpose.

To a Scientist it is a random collection of individual particles having no cohesion and subject to random loading. It is thus fair game for all sorts of his favourite statistical analysis, probability distributions or even joint probability distributions - and in 2 and 3 dimensions. The Scientist usually considers that unless a harbour engineer has at least 10 years of wind and wave recording at his site, he is a fool ever to take the job in the first place.

To a Contractor a rubble mound structure is just one big muck shifting job of pouring endless loads of rubble into the sea. He is cynically amused by the specification which calls for tolerances which he knows cannot be achieved but is comforted by the thought that once in place can rarely be properly inspected and measured. He views the scientist with deep suspicion and wishes that he would, just occasionally, leave the rarified atmosphere of his laboratory and find out how it is really done. He has heard that scientists are somehow concerned with the design which ne believes explains all his problems.

To the Structural Engineer it is a stunningly crude structure, which as it is not based on a Code of Practice is therefore despicable and has no right to stand up anyway.

To the Architect who has designed a perfectly proportioned yacht marina development it is an aesthetic disaster which usually ends up going green and smelling horribly.

To the Harbour Engineer it is rather like his mother-in-law...Great to get away from but occupying a unique place in his heart!

But to be serious... What is the 'State of the Art' in 1983 as far as rubble mound breakwaters are concerned?

It is, I would suggest, a state of uncertainty about many factors affecting all aspects from design to construction to maintenance. Perhaps the situation can best be described by referring to a question taken from a book on Harbour Engineering by Brysson Cunningham and written around 1908.

The quotation in fact refers to mathematical theories of wave formation but is, I believe, just as appropriate in a much wider field. He says, and I quote:

"Many of the theories advanced are merely tentative and lack substantial corroboration; others while generally accepted, are still the subject of speculation and enquiry".

That, to me, sums up much of what is written and said today regarding rubble mound structures.

In the face of the recent serious damage sustained by some major structures I believe we should review the routes we have taken over the last 35 to 40 years in trying to solve the problems. If necessary, we should go right back to what we see to be the basic requirements for this type of structure. We should remind ourselves of what it is we are really trying to do. Much will be gained in this respect if we examine carefully the possible reasons for the failure which has occurred to existing breakwaters. However, it is important in so doing to continually remind ourselves of the 'State of the Art' at the time these structures were designed. Many of the facilities we now take for granted simply did not exist even as recently as 10 to 20 years ago, that is within the period when many of the breakwaters under review were designed.

Reliable wave recorders were not commercially available until 1970 - 13 years ago. Hydraulic model studies were by no means universal much before 1955 and then only with monochromatic waves. It was a further 10-15 years before we really had access to random wave generators, that is some 11-18 years ago. It was also the policy in almost every hydraulic laboratory to assess the damage to the armoured face by counting only the number of units which actually rolled out of position and to consciously ignore those which were rocking. This policy still persisted in some laboratories until a matter of some 5 years ago. The first reports on the possible significance of wave grouping on the stability of armoured slopes were only just reaching engineers a matter of some 6 to 7 years ago. It was virtually impossible to investigate wave overtopping with any degree of confidence before random wave generators were available.

The above is by no means a complete list of what has changed and what has been learnt over the last 10 to 20 years. In this context it is interesting to speculate what would have happened if the presently suggested solutions for the rehabilitation of some of the recently damaged breakwaters had been put forward in the early 1970s. I would venture to suggest that many contractors on receiving the tender drawings and seeing the massive cross section would have submitted alternative designs not all that dissimilar to what was actually built. They could have shown a massive saving in cost, they could have justified their design on the basis of the state of knowledge at that time and the consulting engineer would, most probably, have been heavily criticised for being too cautious. This is by no means intended in any way as a criticism of what has been done then or now; it is in fact a criticism of the 'State of the Art' as it was at that time.

Earlier I suggested that perhaps the time has now come when we must reconsider the whole question of the design and construction of rubble mound breakwaters in the light of what we have learnt particularly over the last 5 to 6 years. There are a number of aspects which I would like to raise today to demonstrate what I mean and I do this in the hope that it will provoke discussion and that perhaps out of these discussions will emerge new approaches and new ideas.

Let us first turn our attention to the question of concrete armour units and begin by briefly tracing the history of their development and remind ourselves of the thinking behind them.

Prior to 1970, the majority of breakwaters were armoured with rock, cubes or rectangular blocks. Before 1932 the size and disposition of these units was a matter of individual judgement based on experience. In 1933 Professor Castro in Spain produced the first formula for estimating the weight of stone used in the armouring of dykes, but the first generally accepted formula, for rock armour, was developed by Professor Irribarren, again in Spain, in 1938. Its world wide acceptance was however, delayed as a result of the Second World War. In 1947 and 48, the Waterways Experiment Station at Vicksburg received authority from the American government to carry out an investigation on 'the Stability of Rubble Mound Breakwaters'. For various reasons, the commencement of this work was delayed until 1951, but then continued through to 1955. The work was carried out under the direction of the then Chief of the Wave Action Section - a certain Mr. R. Y. Hudson. A new formula emerged from this work in which the stability of the rock forming the cover layer was catered for by the introduction of a stability coefficient designated K_d which was established on the bases of model tests with uniform waves. In simple terms, the higher the K_d number the lesser the weight of stone required for a given wave condition.

During this time, in 1950, to be precise, the first of the new breed of specially shaped concrete armour units was being developed in France. It was of course the Tetrapod. Whereas the K_d factor for stone had been established by Hudson at around 3.5 to 4, the new unit had values around 7 to 8.

It was immediately obvious that the use of specially shaped concrete units could result in a reduction in weight of 50% for any given wave condition. The new unit was heavily protected by patents and royalties were charged for its use. As the full significance of this new development became generally known, the quest for further improvement seemed to grip the engineering community and the next 15 years saw the emergence of at least one new shape per year and sometimes as many as 4 or 5 all seeking to increase the K_d value. The highest value attained during this period was for the 'Dolos' unit invented by Eric Merrifield in South Africa around 1963 to 1966. Tests in various laboratories put the K_d value as high

as 33 to 35 and sometimes even higher, although this was viewed with considerable caution, and a lower figure of 25 to 28 was recommended for use in the field. The majority of these units were randomly placed on the breakwater face and in two layers. Other engineers, probably seeking further reduction in cost with equivalent efficiency started to examine the possibility of a single layer of units laid, not randomly, but placed in patterns. The Tribar and Cob are typical of such units. They were subjected to exactly similar model tests as other armour units in order to establish their K_d values, and in the case of Tribars, this was placed at between 12 and 15. In the case of Cob units, which were invented in 1969, it soon became clear that they were functioning in a completely different manner, and it proved virtually impossible to establish a K_d value as the slope remained intact almost regardless of wave height.

It was obvious that the Hudson formula did not apply in this case although it was still thought to be applicable to all other units.

It is worth repeating at this point that the method of defining damage when determining the overall stability of a slope of armour units was to express the number of units which rolled out of position as a percentage of the total number of units on the face. Units which moved less than the principal dimension of that unit, or units which rocked, but did not roll out of position were not included. Prior to starting each test, the model breakwater was subjected to moderate wave action in a bid to induce initial settlement and improve interlock.

There was surprisingly little attention paid during the early period to the consequences of units which rolled or rocked in terms of their structural integrity or the possibility of tension stresses developing, or indeed to the possibility that the Hudson formula might not be appropriate.

It is very difficult to pinpoint exactly when the first doubts began to be expressed about stresses arising in armour units due from wave action. Certainly it was considered around 1970 as steel reinforcement was introduced in Dolos units at Humboldt Bay at that time, but my main recollection is that it was usually discarded on account of cost.

In the case of the applicability of the Hudson formula to predict correctly the behaviour of armour units, the first questioning started around 1973. In 1977, at the Breakwater Symposium in the Isle of Wight, it was certainly a matter of considerable discussion, more particularly as the first results were tabled of the Canadian research on the damaging effect on stability of wave grouping.

Since then, we have had the major failures at Sines, Arzew, Bilbao and San Ciprian, and although tentative moves have been made to re-examine the formula, even to the extent of returning to the Irribarren formula, the present day engineer must still commence his initial design using Hudson and rely for his final design on the use of model studies.

The whole question of the stability of randomly placed armour units is now the subject of much detailed discussion and uncertainty, and I detect a marked reluctance in many areas to the use of special shapes. There seems to be a tendency in the latest designs to revert to the more simple shapes such as the tetrapod, the cube or the Antifer cube, or rectangular blocks.

It would, I think, be argued that in our earlier quest for interlocking units with high porosity and therefore high K_d values, we overlooked or minimised vital factors.

These factors are:

1) With certain armour shapes it is impossible to achieve a 'stable' slope comprising two or more layers of randomly placed units

2) If this is so, it will be equally impossible to prevent rocking. The degree to which this occurs may result in
   a) structural breakage of units
   b) change in the porosity of the total layer
   c) settlement of the units down slope

3) The structural breakdown of the armour unit may be hastened if it suffers impact during placement, or contains hairline cracks as a result of stress induced during casting and curing and which may not be spotted during underwater inspection

4) The size of the unit may be so large that the majority of its strength is taken up in supporting its own weight and there is therefore insufficient reserves to cope with dynamic impacts due to wave action etc.

5) Broken units will, because of their reduced weight, be moved or rolled about by smaller waves, but in so doing will impact with other units and cause further breakage.

6) Armour units are subjected to a form of fatigue loading from the repeated impacts from waves throughout the life of the structure. There is new research evidence to suggest that this can be a further cause for breakage.

7) Even if breakages do not occur, the possibility of settlement down the face remains and this may have serious consequences if the breakwater has a large wave wall. The increased exposure to wave action will certainly result in increased overtopping, but may also bring about the failure of the wall itself.

8) Increased wave run up and overtopping may also occur if the porosity of the armoured face reduces with time.

9) None of the consequences of armour unit breakage outlined already will be reproduced correctly in the model study nor can they be determined mathematically. The physical process involved with wave action on an armoured slope are very poorly understood. There is therefore no means available to the engineer to determine the forces acting on his units. It follows therefore from this that structural design of such units is impossible.

10) Perhaps and most importantly of all, once these armour units are in place on the face of the breakwater, we cease to have any control of their subsequent actions.

Much of what has been said so far is a direct result of the method of placing armour units on the face of the breakwater. The intention is to produce an even distribution of units in all planes usually on the basis of a stated number per square metre of face and in most cases in two layers. The normal procedure is to work out a grid of some sort, for the bottom layer and for the second layer to keep the same grid but offset it in such a way that the units fill the gaps or recesses formed in the first layer. Furthermore, with some armour units even the attitude of block on the slope has to be controlled, i.e. with one leg of the base pointing up the slope in one row and down the slope in the adjacent row.

We have two options for handling these units into position firstly by crane from the crest of the breakwater and secondly by a crane mounted on a floating barge. In both cases we have to contend with wave action. In both cases the armour unit is hanging on a long wire from the jib of the crane. In this position it behaves like a pendulum whose motion is firstly set up by rotation into position in air and then modified or even exaggerated once underwater by wave action, particularly if the site is exposed to persistent swell. To some extent this swinging motion can be damped by the use of guy ropes to the hook of the crane, but only to a limited degree. It is therefore very doubtful if the specified grid position can be maintained with any great degree of accuracy, and the unit is still not actually sitting on the face.

There must always be a hazard at this stage that the swinging mass may collide with units already in place and cause cracking, which is then almost impossible to detect, or in the worst case breakage. This sort of risk is highest when placing the second layer.

Assuming however that we are laying the first layer and that our unit has reached almost to the rock underlayer without mishap, with some multilegged units the act of it first touching the stone causes it to rotate before finally coming to rest. The degree of rotation will depend on the shape and attitude of the unit, the irregularities of the stone underlayer which may be considerable and intentional, or further collision or rolling off an adjacent unit. The chances of this unit being in its required grid position are therefore questionable but will it stay in this position. Remember that we are talking about the first layer, a single layer of units, but the type of unit we are talking about is not intended to be stable in a single layer, it is designed to act in an interlocked or wedged mass. The longer the single layer remains in this situation, the greater is the risk that even quite moderate wave action will displace units.

If one accepts that there may be inaccuracies, perhaps in some cases quite large, in laying the first layer, will this not also be reflected in the placement of the second layer? If, for instance, the units on the first layer are not grouped as close together as they should be, will the second layer unit which occupies the space have too much freedom to move, or conversely, if the first layer units are too close together, is there not a chance that the second layer unit will be perched on top producing an arching effect?

The degree of control and inspection available under water is very limited and the visibility frequently restricted to a matter of metres, as a result the engineer may be very unsure as to what is actually the true situation.

Above water level the degree of control is obviously far better and adjustments can be made by eye if the packing of the units does not appear to be adequate but it is still problematical whether or not the mass of units can be regarded as stable. In fact, there could be a marked difference between the stability of the upper half of the armoured slope above water level to what exists below water level at least during the very early life of the structure. Such a possibility will be more critical on the larger breakwaters where there will be a considerable length of face below the water line.

If we are to continue to depend on this method of protecting rubble mound breakwaters, we must, as a matter of urgency, improve existing construction and inspection processes. At the same time, we must, with equal urgency, explore alternative methods of protection.

One way of achieving a stable slope which retains its original hydraulic characteristics over a long period of time is the use of single layer pattern placed units. I am thinking of the Cob unit, the Shed, the Seebee and the Diode, all of which are now in service. All have excellent potential for long term stability, with high but permanent values of porosity giving increased energy dissipation, low reflection coefficients and much reduced levels of wave run up. Like all new armour units they have their critics who point to difficulties in the design and construction of the toe and at the crest particularly in front of a wave wall, and of the need for very careful trimming of the underlayer immediately below the units. This latter aspect is greatly helped by the need to use a much smaller size of stone in the underlayer than would be the case with many randomly placed armour units and experience in the field has shown no great difficulties in this respect.

With regard to the design of the toe, I believe I am right in saying that in the situations where they have been used, access to the toe was possible during low tidal levels which enabled the engineers to design for solid mass concrete buttresses or vertical sheet piling. If they are to be considered as armouring for a more conventional breakwater, attention will have to be paid to these particular aspects. I do not think they will prove to be insoluble.

I believe that with careful development they will also offer a more predictable solution to the problem of placement on the breakwater face resulting in a higher degree of confidence in the end result. In this respect, a small change in the basic geometry of this type of block may help with the problems at the toe, the crest and the back face armour.

I want to move on now to the question of wave overtopping and crest walls or wave walls.

Research on the subject of wave overtopping shows that increased stability can be expected if some of the energy in the uprushing wave is allowed to pass over the breakwater. In many instances, I believe we give less weight to this concept than we should and whilst it is clearly a matter of degree there are many harbours where this could be tolerated without detrimental effect. Conversely, there are other situations where it is not acceptable, particularly if areas are to be reclaimed behind the breakwaters and used for cargo handling, etc. There are other cases too where it is felt that the very large expenditure associated with the construction of the breakwater should be justified by its use as a berthing facility, perhaps for special cargos of a hazardous nature which have to be kept clear of general port operations. It may, therefore, be required to carry oil pipes or conveyor systems. Finally, of course, it may be necessary to provide access along the breakwater in order to carry out the maintenance of the breakwater itself.

Whatever the reason, if a wave wall has to be provided, the design engineer is ill served by the present state of knowledge in determining his loading. There are no reliable guide lines to help him. He is also without adequate guidelines when dealing with the design of the toe berm and scour apron although these matters are crucial to the stability of the armour layer.

In the matter of the design of the core and rock filter layers much of the existing information is based on the design of well filters subject only to uni-directional flow as set out by Terzaghi. Research has started on this subject but much more needs to be done to study the effect of oscillatory flow in the voids of graded sub-layers. It has to be remembered that the physical hydraulic model cannot help in this respect due to scale effects. Although, provided the scale used is sufficiently large, these scale effects are minimal in the primary under-layer.

However, some assumptions must be made with respect to the size and grading of these various layers prior to the model test and this frequently has to be done in the absence of field surveys, quarry searches or trial blasts. If these assumptions are subsequently proved to be wrong or different or if the characteristics of the stone changes during the construction there is at present no means of adequately assessing this effect on the breakwater's long-term stability. The true porosity of a graded sub-layer also needs research taking into account the variation in the shapes of the rock itself. Whilst it is essential to specify limits, such as the maximum to minimum dimensions measured at right angles being no greater than 1:2 or 1:3, the variety of shapes which can be included within that specification is large.

Much more needs to be known about this matter if we are to prevent the sucking out of rocks and smaller particles from the heart of the breakwater and so induce both settlement and the knock on effects on the stability of all the other elements of the breakwater, including both armour units and wave walls.

There is no time to go into all the geotechnical aspects other than to list some of the main problems.

1) Excessive water pressures acting on slip circles
2) the influence of wave action on soil characteristics, especially pore water pressures in the soil forming the base of the breakwater

3) the need to develop a fully dynamic approach to slope stability under random wave action

4) determination of pore water pressures within the breakwater due to wave action

5) dynamic effects caused by earthquakes, tsunamis and explosions, also the possibilities of liquefaction.

Many of these approaches are based on mathematical modelling and at present depend on what I might term 'the monochromatic wave approach'.

This is said in no way as a criticism bearing in mind the complexities involved, but I would add a word of caution.

We have learnt in the past the possibilities of error when using monochromatic waves as opposed to random waves. Perhaps we may also find that random wave mathematical models will show the same sort of differences as the physical models.

In conclusion, I believe that the tools now available are better than we had 10, 15 or 20 years ago. Our knowledge had advanced but it needed some very serious failures to make us stop and think.

Much more research is needed to reduce the level of risk which all engineers have to accept at present if they choose to be concerned with the design and construction of breakwaters.

Gentlemen, it might be said that the real source of all our troubles is 'The Sea Wave' and I would like to end with a most eloquent description of 'our enemy':

"By far the mightiest of the forces arrayed against the harbour barrier is the sea wave. This mysterious product of wind and water is endowed with tremendous disruptive power. It acts with all the magnificent impulse of a huge battering ram, while, at the same time, it is equipped with the point of a pick and the edge of a wedge. It is, in fact, one of the most complex, the most volatile, the most pertinacious and the most uncomprehensible of natural forces."

(Brysson Cunningham, 1908)

1 Synthesis of design climate

J. A. BATTJES, Delft University of Technology

The synthesis of climate statistics for breakwater design is dealt with, particularly with respect to the waves. Sources considered are wind data and wave data, both visual and instrumental. The need to extract information from all potential sources is stressed. Some methods to be used for this purpose are mentioned, and an indication is given of results which have been obtained. Uncertainties inherent in estimation of climate parameters are discussed, particularly for extreme events.

INTRODUCTION

1. This paper gives an overview of the state of the art in the specification of environmental conditions needed in breakwater design, as far as their effect on the structural behaviour is concerned. The environmental conditions considered to be relevant for this purpose are sea level and sea state, to be treated on a climatic time scale. However, because of space restrictions, aspects of sea level will only be mentioned in passing.

2. The nature of the problems addressed, and the diversity of individual cases, rule out the possibility of a single approach which would be generally applicable. Certain suitable procedures can be sketched or referred to, and certain general principles and inferences from experience can be formulated. This is what has been attemped in this paper.

SEA STATE PARAMETERS

3. Breakwater design procedures usually involve scale model testing to investigate the behaviour of the design-structure under wave attack. This requires an adequate simulation of sea states, including their irregular, random character. The necessity of this procedure is generally accepted and needs no further discussion.

4. There is some controversy about what represents an adequate simulation of random sea states, particularly with respect to wave groups. For the present purpose the information equivalent to that contained in the spectral distribution of wave energy only is considered. This is deemed sufficient in view of the empirical evidence supporting the applicability of the model of independent, uniformly distributed random phases, also for the group statistics (refs. 1, 2).

5. It is common practice to only use long-crested waves for breakwater testing. It seems unlikely indeed that the inclusion of short-crestedness would alter the structural behaviour to a sufficient degree to warrant the considerable extra effort and cost, at least in the case of rubblemound breakwaters. For monolithic types, where the total load on a large element is important, the directional energy distribution is more important, but in that case its effects on the wave-induced loads (a reduction) can be estimated computationally (ref. 3). The use of a generalized parametric representation of the directional spectrum (refs. 4, 5) is sufficient for such purposes. Since moreover the parameters of the directional spectrum are not yet measured on a routine, long-term basis, they will be omitted for the time being from the parameters to be considered in the context of a wave climatology, except for the mean or principal wave propagation direction (θ_s), which can be estimated visually or be correlated with wind direction.

6. As a consequence of the above, only the 1-dimensional frequency spectrum $F(f)$ is considered here (apart from θ_s). This spectrum can in principle have any shape, but in a pure sea (wind-driven waves) it is generally found to have a quasi-equilibrium similarity form which can be described with only two independent spectral parameters, e.g. the peak frequency (f_m) and the total area (E).

7. Instead of the spectral parameters (E, f_m) one can use the significant wave height and period (H_s, T_s), using the empirical approximate relations for sea (ref. 6)

$$H_s \simeq 3.8 \ E^{\frac{1}{2}}, \ T_s \simeq 0.95 \ f_m^{-1} \qquad (1)$$

DATA

8. Observational data can be subdivided into categories according to various criteria:
o according to the variables observed:
  sea surface pressure, winds, waves
o according to method of observation:
  visual vs instrumental
o according to the duration:
  short series (a few years at most) vs long series (several decennia)

Breakwaters—design and construction. Thomas Telford Ltd, London, 1984

9

o according to the sampling interval:
  constant vs random
o according to the location:
  offshore, nearshore, coastal

9. The ideal of having instrumental wave data, collected at or near the required site at constant intervals during a long time span is presently unattainable, and this will remain so for most locations, at least in the foreseeable future. (Routine programs of satellitebased remote sensing may change this on a global scale.)

10. Though each of the data sets actually available will suffer from one or more shortcomings, compared with the ideal sketched above, they nevertheless contain at least some useful information. It is wise therefore to reject none a priori, but to <u>maximize the data base</u>, and to extract the maximum amount of information from the available data for the purpose at hand. Getting the raw data may require search procedures in various files, but this is well worth doing for major projects, considering the enormous costs of underdesign and of overdesign.

11. The maximization of the data base should be both quantitative, referring to coverage in space and time and to variables considered, and qualitative, implying procedures of quality control, intercomparisons, and so on. The extension of the time base is particularly crucial for estimation of extremes. This implies the use of wind data and perhaps visual wave data, which in turn requires procedures such as:
o Validation of visual data against instrumental data.
o Correlation of offshore wind data and coastal wind data.
o Correlation of wave data with wind data.
o Hindcasting waves from wind data.

VALIDATION OF VISUAL DATA
12. Tests of validity of visual data should in general deal with aspects of variability and bias. However, since we are here dealing with the use of long series of visual data in climatic studies, we will consider the aspect of bias only.

Windspeed
13. Under auspices of the WMO, visual estimates of windforce and direction are made and reported by voluntary observing ships en route. The windforce is estimated in terms of Beaufort numbers and converted to a windspeed using a scale which was introduced internationally in 1948. However, subsequent comparisons with measured windvelocities showed a bias, such that low windspeeds were underestimated and high windspeeds were overestimated. A new scale was therefore adopted, but in order to avoid confusion among the observers the rules for reporting the windspeed estimates were not changed. Thus, the new scale must be used in a process of correcting windspeeds reported on the basis of the old scale (ref. 7).

14. Quayle (ref. 8) has made a comprehensive climatic comparison of visually estimated windspeed with measured values, for a number of ocean

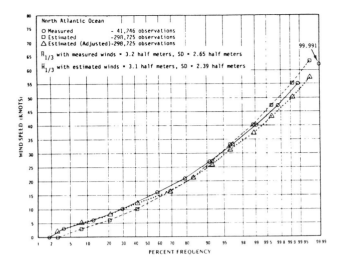

Fig. 1 Measured (o) and visually estimated wind speeds; □: old scale; △ : new scale; (ref. 8).

regions. Both the old scale and the new one were used in the comparison. Fig. 1 gives a result in terms of cumulative frequencies for a region in the North Atlantic. From these data and others (ref. 9) it appears that visually estimated windspeeds can be used with confidence in climatic studies, provided the new scale is used.

Wave height and period
15. According to WMO convention (ref. 7), visual estimates of a characteristic wave height (H_v) and period (T_v) of a wave system (sea and swell (s) considered separately) should be based on mean values of about 15 to 20 well-formed waves of the higher waves in several groups. It is therefore natural, in comparisons with instrumental data, to correlate these visual estimates with mean values of the higher waves in a record, for which purpose usually the upper onethird fraction is taken, which results in the so-called significant wave height (H_s) and period (T_s).

16. Using equal frequency of exceedance as a criterion, Nordenstrøm (ref. 10) has derived the following relationship between visually estimated and recorded wave heights at ocean weather ships in the North Atlantic:

$$H_s = 1.68 \, H_v^{0.75} \qquad (2)$$

where H_s and H_v are expressed in metres. A comparison for a North Sea location has been made by Hogben (ref. 11), showing good agreement between H_v and H_s. More generally, Andrews et al. (ref. 12) state that "it is in fact widely considered that, given large enough samples of data, statistics based on visual observations of wave height are sufficiently reliable for many engineering purposes". A similar view is expressed by Jardine (ref. 13).

17. The reliability of visual wave height data can be enhanced if they are treated jointly with wind data, as proposed by Hogben (ref. 14). This procedure is outlined below. At this stage it suffices to note that the improved estimates of

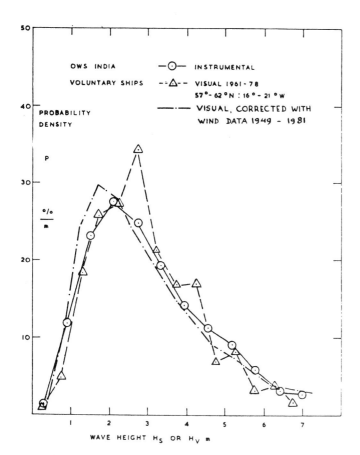

Fig. 2   Measured and visually estimated
wave heights (ref. 16).

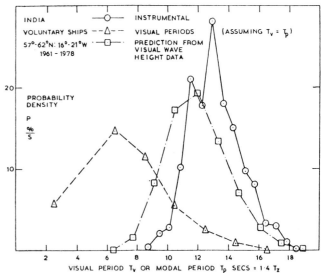

Fig. 3   Measured and visually estimated
wave periods (ref. 15).

visual wave height statistics are in fair agreement with those based on instrumental data, provided the latter are also of sufficiently long
duration (see Fig. 2).

18. The results with respect to the reliability
of visually estimated wave periods seem to be
more divergent than those for the wave heights.

19. Despite large scatter in the data, one can
consider average proportional relations between
T_v and instrumentally determined characteristic
periods. The 1979 ISSC Report (ref. 6) mentions
average values of the ratio T_v/T_m, in which
$T_m \equiv f_m^{-1}$, the period corresponding to the peak-
frequency of the energy spectrum. Results from
three different sources are quoted, for which
the averages of T_v/T_m ranges between 0.89 and
0.94. Comparing these values with eq. 1, and
recalling the motivation for the definition of
significant wave parameters, the 1979 ISSC (ref.
6) concludes "it is recommended that T_v should
be assumed equal to $T_{H_{1/3}}$" (where $T_{H_{1/3}} \equiv T_s$).

20. The view expressed in the quotation just
given is not held unanimously. Fang and Hogben
(ref. 15) e.g. state that "broadly it may be said
that visual wave height statistics are adequate
for many engineering purposes provided they are
based on reasonably large samples of observations
but visual wave period statistics are generally
rather unrealiable". In fact, they propose not
to use the visually estimated periods at all, but
to use a generalized conditional p.d.f. of wave
periods, given the waveheights, with parameters
estimated from instrumental data.

21. The use of joint statistics of significant
wave height and period in processes of quality
control and parameter estimation is recommendable
since the two variables are in fact correlated,
and also because the joint statistics are needed
anyway in applications, not just the marginal
distributions.

22. Fang and Hogben (ref. 15) have approximated
the joint probability density function (p.d.f.)
of H_s and the mean zero-crossing period, T_z, with
a two-dimensional log-normal distribution, modified so as to incorporate a skewness of the p.d.
f. of $\ln H_s$. Using 23 instrumental data sets,
they derived empirical relations between the
mean and standard deviation of $\ln T_z$ and of the
coefficient of correlation between $\ln T_z$ and
$\ln H_s$, as functions of the overall-mean value of
H_s. These results were used subsequently to estimate period statistics from visual wave height
data. An example of the results so obtained is
given in Fig. 3. This figure is typical of the
set shown in refs. 15 and 16. It indicates the
poor quality of the raw visual period data as
well as the considerable improvement obtained
through the use of the joint p.d.f. of heights
and periods.

CORRELATION OF WIND- AND WAVE-DATA
23. Hogben (ref. 14) has proposed and described
a method of using the joint probability distribution of wind- and wave parameters in climate
studies. In Hogben's method, the conditional
p.d.f. $p(H_s|U)$ of H_s, given a certain value of
the windspeed U, is estimated from paired observations of H_s and U. ("Paired is roughly equivalent to "simultaneous", but it is also possible
to pair values of H_s with those of U which occurred one hour or a few hours earlier.) This estimate can subsequently be combined with the long-
term p.d.f. $p(U)$ of U, based on long-term observations of longer duration, to estimate the p.d.
f. of H_s. Hogben assumes a two-parameter gamma
type p.d.f. for $p(H_s|U)$, and gives empirical re-
lations for the conditional mean and standard deviation of H_s as a function of U. The values for

U = 0 are considered as swell.

24. There are several modes in which the approach sketched above can be used:

(a) The model can be fitted to a data set of paired (H_s, U) values, with the advantages of obtaining a smoothed distribution (compared to the raw frequencies of observation) represented by only a few numerical parameters (whose values are estimated from the data).

(b) In conjunction with longer-duration wind velocity data, the model can be used to give improved estimates of the long-term p.d.f. of H_s, using the parameter values estimated as in (a).

(c) If at the study site wind velocity data are available but no wave data, the model can be used as a predictor of the p.d.f. of H_s, using parameters values estimated from other locations.

An example of results of mode (b) is given in Fig. 2, showing the smoother distribution of the revised values compared with the raw ones, and a better correspondence with instrumental data.

25. With respect to mode (c) above, it is pointed out that the parameters should be expected to depend on location, because of different weather pattern and/or different degrees of exposure. The use of Hogbens's model in this predictor-mode is therefore not advisable unless one is dealing with a locality where the meteorological conditions and the exposure are roughly comparable to those for which the parameter values were estimated. Even then the variations of the parameters between stations can be significant, as is evident from an inspection of the figures in ref. 14.

26. Hogben's model of correlating wave data with wind data as sketched above does not consider wave periods, but these could of course be treated in a similar fashion in terms of $p(T_s|U)$, with the same advantages. One can similarly use a windspeed-dependent conditional joint p.d.f. $p(H_s, T_s|U)$, which perhaps can be modelled after the marginal joint p.d.f. $p(H_s, T_s)$ as used by Fang and Hogben (ref. 15), discussed earlier.

27. A possibly useful alternative or addition to the above is to consider a so-called universal relation between total wave energy E (actually, E is the variance of the surface elevation), the peak frequency of the energy spectrum, and the windspeed U, for wind-driven waves. This relation is "universal" in the sense that it holds regardless of the growth stage of the waves, and that it is not restricted to constant windfields with well-defined fetches or durations. Reference is made to the following section on wave hindcasting, and in particular to eq. 4, which on substitution of eq. 1 can be expressed in terms of H_s and T_s as follows:

$$H_s^2 T_s^{-3} \simeq 1.15 \times 10^{-4} g \, U \qquad (3)$$

This relation could be used in conjunction with joint statistics of (U, H_s) (as obtained in Hogben's model) to derive joint statistics of (H_s, T_s).

WAVE HINDCASTING

28. Wave hindcasting is the process of estimating wave conditions from historical wind data. It plays a crucial role in most major climate studies in which the statistics of extreme events are needed.

29. Methods of wave hindcasting and of wave forecasting both comprise a wind model and a wave model. What distinguishes hindcasting from forecasting is that the former uses historical wind-data, whereas the latter is based on forecast winds. Therefore, the wind model for hindcasting can use as input not only synoptic meteorological information available at the time of the event presently being hindcast, such as historical weather charts, but also wind observations which have since been collected and archived. The full use of this potentially valuable additional observational material requires a search procedure in files and archives to retrieve the data, as well as a re-analysis of the surface pressure data, with the observed winds now serving as a constraint. For a description of this procedure, and of wind models more generally, including intercomparisons and validation studies, see refs. 9 and 17.

Numerical deep water wave models

30. Numerical wave models considered here are designed to calculate the space-time (x,y,t) evolution of the two-dimensional energy spectrum $F(f,\theta)$ in response to a given windfield $\vec{U}(x,y,t)$. They are based on a differential form of the spectral energy balance equation which besides linear transport terms has a source function $S(f,\theta)$ representing the net effects of atmospheric input (S_{in}), dissipation $(S_{ds}$, mainly due to whitecapping) and (in some) nonlinear wave-wave interactions (S_{nl}).

31. Following ref. 18, the following three principal categories of models are distinguished, viz. those based on:

(a) discrete spectral components without nonlinear wave-wave interactions, or

(b) discrete spectral components with nonlinear wave-wave interactions, or

(c) a parametric spectral representation, used mostly in hybrid form with (a).

Individual models will not be listed here. Instead, reference is made to published resumes such as in refs. 18 through 23.

32. The principal change in the development of the various categories of models was the inclusion of nonlinear wave-wave interactions, derived theoretically by Hasselmann (see ref. 24 for a resume) and shown to exist empirically in the JONSWAP project (ref. 25). These appeared to be the most significant cause of growth of the low-frequency flank of the windsea spectrum, and of the corresponding lowering of the peak frequency. Thus, whereas the tuning of the earlier, linear models (category (a)) to observed growth rates required coefficients in S_{in} far in excess (one or more orders of magnitude) of the ones according to the Phillips-Miles theory (which have since been confirmed empirically from direct measurements of energy transfer across the air-water interface, see ref. 26), this is not the

case if S_{nl} is included properly.

34. The nonlinear interactions were found to have a shape-stabilizing influence on the wind-sea spectrum, causing it to tend rapidly towards a universal equilibrium shape, characterized by a few parameters only. Thus the idea of a parametric windsea model suggested itself. It was developed formally from the full spectral energy balance in ref. 27, but earlier, more intuitive spectral models had already used similar ideas (refs. 28, 29).

35. Parametric models can be used only for the windsea part of the spectrum, because the continuing energy input from the wind is needed to maintain the spectral energy at the levels required to make the nonlinear transfers dominate. In applications to open water, they must be used in conjunction with a discrete spectral model for the representation of swell, giving rise to hybrid models. These utilize an off-on switch between either a windsea mode or a swell mode, whereas actually there is a gradual transition. The difficulties associated with this are avoided in models which combine a discrete representation of the spectrum with a parameterisation of the nonlinear transfers.

36. Several comparisons of operational hindcasts or forecasts with measurements have been published (see refs. 21, 30, 31, 32). Some generalized conclusions from these can be stated as follows:
(i) The development of new wave models or modification of existing ones, in response to the advances made in understanding of the physics of wave generation and propagation, has not (yet) given rise to a correspondingly better performance of these models operationally. At least two factors contribute to this paradoxical situation. One is the limited quality of the wind input. Another is the fact that models with less realistic modelling of the physics compensate for this with more empiricism. It is not surprising that such models, tuned so as to simulate not only observed growth rates but a condition of saturation as well, are capable of fairly realistic predictions in the more or less common situations similar to those for which they were tuned. It is to be expected however that the physically better founded models will perform better in more demanding conditions.
(ii) For the majority of the models, the overall r.m.s. relative difference between hindcast and observed H_s is in the range 0.2 to 0.3 (refs. 21, 22), but the error in the prediction of extreme conditions is smaller by approximately a factor 2 (apart from possible phase errors). An example is shown in Fig. 4 where results of a hybrid model (NORSWAM) and of a discrete model with nonlinear coupling (METOFFICE) are compared to observed H_s-values at the Brent B location.
(iii) Despite the limitations metioned above, hindcast models form a useful and often indispensable tool in climate syntheses. It appears that thay can be tuned so as to have virtually no bias (systematic error). The large random error obtained during calm or moderate conditions is relatively unimportant in climate studies because of the large sample size. For extreme conditions the number of cases is far less but there the random error per prediction is smaller.

Parameter relationships for windsea-spectrum in deep water

37. An observed feature of the frequency spectrum of wind-driven waves, explained theoretically in ref. 27, is its shape-invariance in a wide range of conditions of variable winds. To first order all dimensionless spectral parameters are found to be dependent on a single growth stage parameter only. This has in ref. 27 been worked out for a JONSWAP-shaped spectrum, scaled with g (the gravitational acceleration) and the windspeed U at some 10 m height above sea level (U may be variable). The dimensionless peak frequency $\tilde{f}_m = Uf_m/g$ was used as the growth stage parameter ($\tilde{f}_m > 0.13$ for sea), and other spectral parameters were expressed as functions of $\tilde{f}_m$. The same has been done by Mitsuyasu et al. (ref. 33), using other data. We shall here give a resume of the relations as given by Mitsuyasu et al. since these appear to fit slightly better to the combined data of ref. 27 and ref. 33 than the relations presented in ref. 27.

38. The relation between $\tilde{f}_m$ and the dimensionless energy (variance) $\tilde{E}_m \equiv g^2 E/U^4$ is given in ref. 33 as

$$\tilde{E} \tilde{f}_m^3 = const \simeq 6.84 \times 10^{-6} \qquad (4)$$

39. The JONSWAP-spectrum can be written as

$$\qquad\qquad\qquad\qquad\qquad (5)$$

$$F(f) = (2\pi)^{-4} \alpha g^2 f^{-5} \exp\left\{-\frac{5}{4}\left(\frac{f}{f_m}\right)^{-4}\right\} \gamma^{\exp\{-\frac{1}{2}(\frac{f-f_m}{\sigma f_m})^2\}}$$

Assuming the mean JONSWAP values $\sigma = 0.07$ for $f < f_m$ and $\sigma = 0.09$ for $f > f_m$, Mitsuyasu et al. give the following equations for α and γ:

$$\alpha = 3.26 \times 10^{-2} \tilde{f}_m^{6/7} \qquad (6)$$

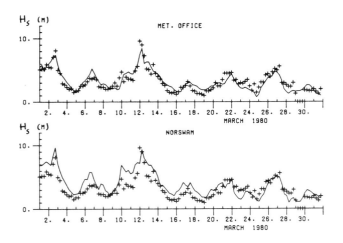

Fig. 4 Measured (+) and hindcast (—) significant wave height at Brent B (ref. 31).

$$\gamma = 4.42 \; \tilde{f}_m^{\;3/7} \qquad (7)$$

40. It is stressed that the windsea spectrum relationships given above are not only for conditions of uniform, stationary winds but also for variable winds, provided there is sufficient dynamical interaction between waves and wind. They can be used in conjunction with joint statistics of (H_s, U), as modelled by Hogben for instance (ref. 14), to estimate long-term statistics of spectral parameters for sea (or (H_s, T_s), as pointed out in paragraph 27).

Numerical shallow water models
41. Incorporation of finite-depth effects in wave models requires not only an adaptation of the terms used in deep-water models, but inclusion of new ones as well. Moreover, prediction of mean sea level (tide, stormsurge, etc) becomes an integral part of the problem, whereas currents also tend to become more important than in deep water. For a resume, see refs. 20 and 34.

42. The most important additional effects needed in shallow-water wave models are those of depth-induced energy redistribution (refraction/shoaling/scattering) as well as dissipation (percolation, bottom motion, bed friction). Also, the non-linear interactions are enhanced relative to their deep-water values.

43. Estimation of the bottom-induced energy dissipation requires knowledge of the relevant parameters of the sea bed material. Uncertainties in these values (and in the modelling) make it necessary to tune each model to the local situation, even more than in the case of deep-water models.

44. A topic of importance is the saturation range of the wavenumber or frequency spectrum. Extending Phillips' arguments for the deep-water case (ref. 35), Kitaigorodskii et al. (ref. 36) have derived a depth-dependent saturation spectrum which in shallow water can be approximated as

$$F_s(\omega) = \tfrac{1}{2}\alpha \; gh \; \omega^{-3} \; \text{for } \omega \equiv 2\pi f \; \stackrel{\sim}{<} \; (g/h)^{\tfrac{1}{2}} \qquad (8)$$

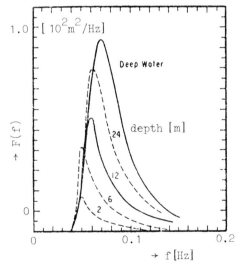

Fig.5   Calculated spectra for shoaling waves with Kitaigorodskii saturation level, $\alpha = $ const. (ref. 34).

The ω^{-3} dependence of the saturation spectrum in shallow water has also been derived by Thornton (ref. 37), using different arguments. Thornton also presents evidence supporting eq. 8.

45. The result of application of the Kitaigorodskii et al. limiting spectrum to a shoaling wavefield is shown in Fig. 5, which clearly shows the lowering of the spectral levels with decreasing depth. The spectra shown were calculated by linear shoaling, then modified to accomodate the saturation spectrum of Kitaigorodskii et al.

46. Most shallow-water models use a discrete spectral representation. Depth dependences of propagation, source functions and equilibrium spectral shape have recently been incorporated in an extended hybrid wave model (parametric sea, discrete swell) by Gunther and Rosenthal (ref. 38).

STATISTICAL ANALYSIS AND PREDICTION
Purpose
47. On climate scale, short-term environmental parameters such as windspeed (U) and direction (θ_w), sea level (ζ), and seastate parameters (e.g. H_s, T_s, θ_s), are considered as a multi-dimensional random process in time (for a given locality). A multitude of aspects of its probability structure can in principle be estimated from observations. Which aspects should be chosen for analysis depends on the purpose for which they are going to be used.

48. It is assumed here that the climatic data are required as input in a process of economic optimization of the design of a breakwater with a given, planned service lifetime, since that is a meaningful and rational procedure. This process will include the estimation of a loss function representing economic losses due to structural damage or failure, damage to auxiliary equipment, associated downtime, etc., for a variety of design alternatives. In this procedure the climatic data analysis should be aimed at the estimation of those aspects needed to evaluate the loss function.

49. Due to the relatively high direct and indirect costs of damage, the economic optimum design of breakwaters is such that the structure is allowed to suffer significant damage only during exceptional environmental conditions. If these occur then the damage is cumulative. These two points imply that the design climate synthesis should be aimed at estimation of probabilities of occurrence of extreme conditions with associated durations.

50. In the procedure sketched above, a "design sea state" or a "design wave height" cannot be defined. Therefore, a "return period" associated with such condition plays no role either.

Approach
51. All procedures of assessing probabilities of future environmental conditions necessarily rest on the a priori assumption of climatic stationarity on the time scales considered. This implies that the probability structure of the environmen-

tal parameters in the (future) time interval of prognosis is assumed to be the same as it was in the (past) time interval of diagnosis. This assumption seems to be the most rational one among possible alternatives when there are no obvious physical limitations (e.g. depth) coming into play. Nevertheless, here is a fundamental uncertainty which can neither be avoided nor assessed.

52. The estimation of extreme conditions, outside the range of observed values, relies necessarily on extrapolation, for which purpose an analytical distribution function is fitted to the data. The following two types of statistical populations of the environmental parameters can conceptually be considered:
(a) the values defined for continuous time; or
(b) the values associated with discrete events (such as storms).
Note: although the population in (a) is defined for continuous time, it is usually sampled at discrete times, e.g. at constant intervals of 3 hours, or at arbitrary instants. It is necessary to be aware of this distinction.

53. The joint p.d.f. of the population of parameters at an arbitrary time (case a) is not a sufficient environmental input for the estimation of the losses to be expected in the planned lifetime of the breakwater. It needs to be supplemented with information about persistence, such as durations of or recurrence intervals between (independent) occurrences of extreme conditions. Such information can only be obtained from analysis of the parameter time histories. (Contrary to frequent practice, recurrence intervals between independent physical events cannot be based on the value of an arbitrarily chosen sampling interval such as 3 hours.)

54. The optimization design procedure must include all multi-parameter environmental conditions contributing to the loss function. Therefore, the discrete-event approach (b) must in this application be based on a criterion of exceedance-of-threshold (e.g., H_s, T_s simultaneously above some minimum). An approach using monthly or annual or other periodic maxima, which can be applied in estimating extremes of a single parameter such as H_s, is less appropriate here because the loss is not determined by any single parameter.

55. An advantage of the approach (b) based on discrete events is that these can usually be assumed to be independent, as required in common statistical methods of estimating sampling variability.

56. A potential disadvantage of (b), compared to the continuous-time approach (assuming that data for the latter are available), is the partial waste of data. However, this is offset to a large extent by the fact that the conditions corresponding to the "wasted" data, i.e. those below the threshold used for defining the discrete events, do not contribute to the damage. Conversely, if the continuous-time data are not already available then it does not pay to generate them. Hindcasting of discrete events is then the most economical approach.

Analysis
57. One can distinguish the following steps in estimating the probability distribution of a population from a finite sample:
(1) Hypothesis about the type of distribution of the population.
(2) Parameter estimation ("fitting" the distribution to the data) by some chosen method.
(3) Acceptance or rejection of the hypothesis that the sample was from a population distribution as assumed, with the parameters as estimated, using some chosen criterion.
(4) Repetition of the preceding cycle for a number of different hypotheses about the population distribution.
(5) Ranking the accepted distributions according to some chosen measure of the likelihood that they would be the population distribution, and choosing a favoured one.
These steps will not be detailed here. They have been well documented e.g. in ref. 39 for the one-parameter case of the maximum H_s per storm. In the present application, we have to deal with a multi-parameter population, which requires the use of joint p.d.f.'s. This does not change the essence of the method, but it requires more automated procedures and less reliance on visual methods. For instance, the parameters must be estimated numerically, using a method of least squares, or of maximum likelihood, etc.

58. The joint p.d.f. $p(H_s, T_s, \theta_s)$ can be factorized into the marginal p.d.f. $p(\theta_s)$ and the conditional joint p.d.f. $p(H_s, T_s | \theta_s)$. It is usually sufficient to represent $p(\theta_s)$ in discrete form (wave rose, tables) without analytical fit, since there is no need to extrapolate outside the range of observed values. For $p(H_s, T_s | \theta_s)$, various analytical functions can be tried, such as the two-dimensional log-normal type (ref. 15; see also paragraph 22) or the two-dimensional Weibull type. (The latter has been applied successfully to the heights and periods of individual waves in ref. 40.) An alternative is to work with $p(H_s)$ and $p(T_s | H_s, \theta_s)$ as two separate one-dimensional p.d.f.'s, but in that case the parameters of $p(T_s | H_s, \theta_s)$ should be estimated as an analytical function of (H_s, θ_s) to allow extrapolation. This procedure has been applied to $p(H_s, T_z)$ in ref. 41, where $p(H_s)$ was taken of the Weibull type and $p(T_z | H_s)$ of the log-normal type, with satisfactory results.

59. Statistics of sea state duration can be estimated from time series of local sea state parameters, if such data are available. In refs. 42 and 43, the results for some locations in the Norwegian Sea and the North Sea have been interpreted and presented in a form permitting some generalization to other locations with similar weathercycles though not necessarily of similar intensity.

60. In breakwater studies, sea level (ζ) is an important additional parameter, not only because it determines the elevation of wave attack on a breakwater but also because of its influence on the incident waves. If the data permit, one can attempt to include it in a statistical analysis. Otherwise, or in addition, one can use wind sta-

tistics in conjunction with relations between wind, wind set-up and windwaves to estimate $p(\zeta, H_s, T_s | \theta_s)$. Such procedure has been applied in a design climate study for the stormsurge barrier in the Oosterschelde estuary in The Netherlands (ref. 44).

Prediction
61. The results of analyses such as sketched above can be used to assess the p.d.f. of damages to be incurred during the planned lifetime of the structure. This requires extrapolation of the accepted distribution(s) of the environmental parameters. The extreme events are usually assumed to occur as a Poisson-process in time, which is based on a priori arguments and supported to some extent empirically (refs. 45, 46).

UNCERTAINTY ASSESSMENT
62. There are several sources of uncertainty in the procedure referred to above. These include:
(1) assumption of climatic stationarity;
(2) errors in data (calibration, sampling variability, analysis, modelling);
(3) uncertainty in type of population distribution;
(4) uncertainty in parameter values (sampling variability);
(5) uncertainty in damage function (scale effects).
It is important to quantify these as much as possible. This is possible to some extent for categories (2), (4) and (5), but not at all for (1), whereas with respect to (3) at least a sensitivity check can be and should be carried out.

63. Concerning (4), ref. 47 gives analytical expressions from which the sampling variability in the estimated parameter values of the frequently used one-dimensional Weibull p.d.f. can be evaluated. If analytical approximations fail or become impractical, numerical simulation studies can be used. This has e.g. been done in refs. 39 and 48. In the latter, sequences of log-normally distributed annual maxima of H_s were generated, contaminated in varying degrees with uncertainties of category (2). The sampling distribution of the estimated value of H_s for a number of sample sizes and return periods was determined. An interesting point in the results was that in all cases tested there was a positive bias (overestimation of H_s for given return period, on the average), increasing to a very substantial level with relative degree of contamination. (It is noted here that the population p.d.f. was assumed to be known; if this is not so then a negative bias would seem to be possible as well.)

64. Quantitative assessments of uncertainties, as in ref. 48 or by similar methods, should be an integral part of a climate synthesis program. Sources of uncertainty at all levels of the analysis should be taken into account, and transformed into statements about the limited statistical reliability of one's final estimates. After all, the uncertainties are there, unavoidably, and it is more realistic and in line with responsible professionalism to be open about them than it is to ignore them and to pretend a degree of certainty which in fact is quite il-

lusory. An explicit, quantitative treatment of uncertainties is also a prerequisite for a meaningful risk analysis.

REFERENCES
1. RYE H. and LERVIK E. Wave grouping studied by means of correlation techniques. Proc. Symp. on Hydrodynamics in Ocean Engineering, Trondheim, 1881, Vol. 1, 25-48.
2. GODA Y. Analysis of wave grouping and spectra of long-travelled swell. To appear as Report of the Port and Harbour Res. Inst., 22, 3, March 1983.
3. BATTJES J.A. Effects of short-crestedness on wave loads on long structures. Appl. Oc. Res., 4, 3, 1982, 165-172.
4. MITSUYASU H. et al. Observations of the directional spectrum of ocean waves using a cloverleaf buoy. J. Phys. Oc., 5, 1975, 750-760.
5. HASSELMANN D.E., DUNCKEL M. and EWING J.A. Directional wave spectra observed during JONSWAP 1973. J. Phys. Oc., 10, 1980, 1264-1280.
6. ISSC Committee I.1. Environmental Conditions, Proc. 7th. Int. Ship Structures Congress, Paris, Vol. 1.
7. WMO. Handbook on wave analysis and forecasting, Rep. 446, 1976.
8. QUAYLE R.G. Climatic comparisons of estimated and measured winds from ships. J. Appl. Met., 19, 1980, 143-156.
9. CARDONE V.J. Windfields for wave models. Transcript of workshop on wind-wave hindcasting and forecasting models, NOAA, Gaithersburg, Md, 1980, 33-80.
10. NORDENSTRØM N. Relationships between visually estimated and theoretical wave heights and periods. Rept. 69-22-S, DNV, 1969.
11. HOGBEN N. Environmental parameters. Proc. Symp. Ocean Eng., Joint Oceanographic Assembly, Edingburgh. Published by ECOR, 1976, 3-18.
12. ANDREWS K.S., DACUNHA N.M.C. and HOGBEN N. Wave climate synthesis, Draft, NMI, 1982.
13. JARDINE T.P. The reliability of visually observed wave heights. Coastal Eng., 3, 1, 1977, 33-38.
14. HOGBEN N. Wave climate synthesis for engineering purposes. NMI, R103, 1981.
15. FANG Z.S. and HOGBEN N. Analysis and prediction of long-term probability distributions of wave heights and periods. NMI, R146, 1982.
16. HOGBEN N. Wave climate synthesis: comparisons of visual and instrumental data for validation of the "NMIMET" program. NMI, 1982.
17. CARDONE V.J. et al. Error characteristics of extratropical-storm wind fields specified from historical data. J. Petr. Techn. May 1980, 872-880.
18. SWAMP (Sea Wave Modelling Project), Report, 2nd draft, July 1982.
19. EWING J.A. Numerical models and their use in hindcasting wave climate and extreme value wave heights. Proc. Int. Conf. Sea Climatology, Paris, 1979, 159-178.
20. FAVRE A. and HASSELMANN K. (eds.). Turbulent fluxes through the sea surface, wave dynamics, and prediction. Plenum Press, 1978.
21. HOLTHUIJSEN L.H. Methoden voor golfvoorspelling (Methods for wave forecasting; in Dutch). TAW, The Hague, 1980.
22. HOLTHUIJSEN L.H. An engineer's review of wave forecasting and hindcasting techniques. To appear in Proc. Offshore Goteborg, 1983.

23. YOUNG I.R. and SOBEY R.J. The numerical prediction of tropical cyclone wind-waves. James Cook Univ., Eng. Res. Bull, CS20, 1981.

24. HASSELMANN K. Weak-interaction theory of ocean waves. In: Basic development in fluid dynamics, 2, Ac. Press. New York, 1968, 117-182.

25. HASSELMANN K. et al. Measurements of wind-wave growth and swell decay during the Joint North Sea Wave Project (JONSWAP). Deutsche Hydrographische Zeitschrift, A, 8, 12, 1973, 1-95.

26. SNYDER R.L. et al. Array measurements of atmospheric pressure fluctuations above surface gravity waves. J. Fluid Mech., 102, 1981, 1-59.

27. HASSELMANN K. et al. A parametric wave prediction model. J. Phys. Oc., 6, 1976, 200-228.

28. HAUG O. A numerical model for prediction of sea and swell. The Norw. Met. Inst., Met. Ann., 5, 1968, 139-161.

29. SANDERS J. A growth-stage scaling model for the wind-driven sea. Deutsche Hydrografische Zeitschrift, 29, 1976, 136-161.

30. RESIO D.T. and VINCENT C.L. Model verification with observed data, Report 2 of a series, U.S. Army Waterways Experiment Station, Misc. Paper H-77-9, 1978.

31. EWING J.A. et al. Comparison of the Meteorelogical Office and NORSWAM wave models with measured wave data collected during March 1980. IOS Report No. 127, 1981 (Unpublished manuscript).

32. BOUWS E. et al. An evaluation of the KNMI operational wave model GONO for the period Oct. 1980 - April 1981. KNMI Techn. Rep. 11, 1982.

33. MITSUYASU H. et al. Observation of the power spectrum of ocean waves using a cloverleaf buoy. J. Phys. Oc., 10, 1980, 286-296.

35. VINCENT C.L. Shallow-water wave modeling. Preprint, First Int'l. Conf. Meteor. and Air/Sea Interaction of the Coastal Zone, Am. Met. Soc., The Hague, 1981, 87-95.

35. PHILLIPS O.M. The equilibrium range in the spectrum of wind-generated waves. J. Fluid Mech., 4, 1958, 426-434.

36. KITAIGORODSKII S.A. et al. On Phillips' theory of equilibrium range in the spectra of wind-generated gravity waves. J. Phys. Oc., 5, 1975, 410-420.

37. THORNTON E.B. Rederivation of the saturation range in the frequency spectrum of wind-generated gravity waves. J. Phys. Oc. 7, 1977, 137-140.

38. GUNTHER H. and ROSENTHAL W. Hindcast of seastate during MARSEN 79. Abstract, First Int'l. Conf. Meteor. and Air/Sea Interaction of the Coastal Zone, Am. Met. Soc., The Hague, 1981, 401.

39. PETRUASKAS C. and AAGAARD P.M. Extrapolation of historical storm data for estimating design wave heights. Proc. OTC Houston, 1970, Vol. I, 409-420.

40. KIMURA A. Joint distribution of the wave heights and periods of random sea waves. Coastal Eng. in Japan, 24, 1981, 77-92.

41. SIGBJÖRNSSON R. et al. Estimation of the joint distribution of significant wave height and average wave period. Report STF71-A76041, SINTEF, Trondheim, 1976.

42. VIK I. and HOUMB O.G. On the duration of sea state. In: Safety of structures under dynamic loading. Tapir Publishers, Trondheim, 1978, Vol. 2, 705-712.

43. GRAHAM C. The parameterisation and prediction of wave height and wind speed persistence statistics for oil industry operational planning purposes. Coast. Eng., 6, 4, 1982, 303-329.

44. VRIJLING J.K. and BRUINSMA J. Hydraulic boundary conditions. In: Hydraulic aspects of coastal structures. Delft Un. Press. 1980, Part I, 109-133.

45. RUSSELL L. and SCHUELLER G. Probabilistic models for Texas Gulf Coast Hurricane Occurrences. OTC, Houston, 1971, 177-190.

46. HOUMB O.G. On the duration of storms in the North Sea. Proc. 1st Conf. Port and Oc. Eng. under Arctic Conditions, Trondheim, Vol. 1, 1971, 423-439.

47. LAWLESS J.F. Approximation to confidence intervals for parameters in the extreme value and Weibull distributions. Biometrika, 61, 1974, 123-129.

48. EARLE M.D. and BAER L. Effects of uncertainties on extreme wave heights. Proc. ASCE, WW4, 1982, 456-478.

2 Specification of construction climate

M. A. MESTA, BSc, MSc, MSE, MSCE, Sezai Turkes-Feyzi Akkaya Co Inc and Massachusetts Institute of Technology

Parameters of design briefly referred in. Major sources of uncertainties existing in the design - construction process are indicated. Need for improved methodology in design analysed with emphasis over the type and wide range of information for construction as well the essence of real data and its capture via prototype observation. Construction climate and the importance of incorporation of construction experience and judgement as well as level of risk into design procedures discussed.

INTRODUCTION

1. The successful Design and Construction of marine structures is highly dependent on a complete knowledge of the wave climate, much more complete than the " Wave Height " and the associated " Period ".

2. Some of the necessary additional parameters from designers point are; wave steepness, joint distribution of wave heights and periods, the wave direction, the crest length of waves, wave grouping and the spectral shape. (Ref.1)

3. An International agreement must be reached on the definition of such additional parameters needed for the proper representation of the physical conditions in nature at the model studies.

4. To determine these parameters and their degree of relevance over the design of a Breakwater; Prototype behaviour - with particular emphasis over their complex interaction between waves as well as failure modes - should be closely observed, the real data acting upon them should be extensively captured and comprehensively analysed. Only after that Design Procedures could be improved, testing of such structures - where nonlinear effects are very important - could be carried successfully and the reliability over their hydraulic models could be increased.

5. In the codification and regulation aspect, we should state and request from every country to establish meteorological and hydrological stations capable of observing all meteorological parameters, including data for operational and critical design criteria. Normally wind, wave, humidity, water level-tide, temperature, drift of ice, and specific combinations of them will be of increasing importance. Academic institutions should clarify the differences in such definitions preventing chaos in the judgements over certain data.

6. The shaken " Reliability " or " Safety " of the methodologies followed in the Design of structures indicating failures or suffering damages, set forth the extent of UNCERTAINTIES

involved not only between the " Data - Conceptual Design - Final Design - Model testing " phases; but as much severely among the " Model testing - Construction - Prototype Observation - Prototype Evaluation (Review) - Performance Monitoring - Operation "phases as well.

7. The following presentation will trace some of these factors and consider suggested definitions or concepts aimed at improving the philosophy of the Design of Rubble Mound Breakwaters.

DESIGNER VERSUS CONSTRUCTOR

8. The Designer delivers a design and a tender ready for bidding. The Contractor delivers the final product. Design goes through phases of data collection ranging from environmental to transportation and economic analysis of the project as a whole. While most of this is not the task of the contractor his work depends heavily upon it while the risks of their incompleteness would also be transferred to him.

9. Designers have been preferring the easy way and switch to analytical computations instead of insisting establishment of the data collection systems, by simply stating " Since observations stretching over a sufficiently long period of time are not available ...", thus even today we feel the consequences.

10. The Designer is aware of the inadequacies in data, however, presents his information saying " .. is supplied solely for assisting the tenderer " or " .. in good faith and without warranties to accuracy and completeness! ". He, through clauses as " Inspection of site ", will still hold the contractor responsible for the collection of all necessary information within the limited tender period which is one fifth or less the design period.

11. Specifications should convey information about the length of the data base provided since all climatological practice shows that it is necessary to have series of at least thirty years in order to get a statistical control of the material. (Ref.2)

Breakwaters—design and construction. Thomas Telford Ltd, London, 1984

19

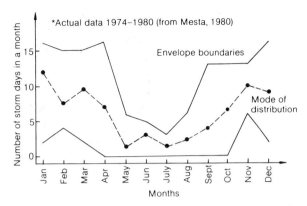

Fig. 1. Periodicity of stormy days on a yearly basis at a jobsite with seven years actual data

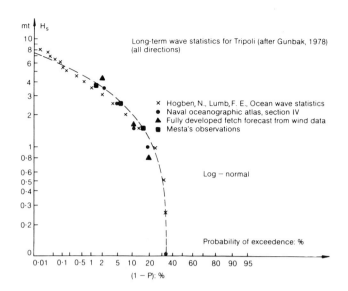

Fig. 2. Visual observations against theoretical long term wave statistics

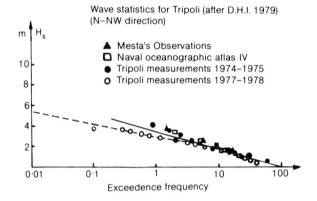

Fig. 3. Visual observations against actual wave measurements at site

Fig. 4. Water jets through holes pointing to pressure distribution at a wave wall

12. Distribution functions for significant wave height, viewed in the exponential format, have indicated that; one complete year of data at six observations per day appears to give a reliable wave height distribution function up to 1% level of occurrence. (Ref. 3) If this is the case together with the fact that the Designer spends in the average over one year yet transfers all the risks of inadequate information to the construction group via contract documents, should be viewed as rather unfair.

13. Due to uncertainties, a marine decision maker finds that the set of decisions to which he can usefully apply quantitative analysis is quite limited. The resulting attitude is concisely expressed in the sometimes skeptical, sometimes derogatory, and sometimes plaintive querry; " HOW CAN YOU ANALYSE WHAT YOU CANNOT PREDICT ?"

CONCEPTS

14. Concepts which are difficult to define become even more difficult to measure.

15. Wind: Both in locations with Arctic or Desertic climates, a very rapid heating and intensive instabilization of the air masses take place by heating of the lowest layers of air, which lead to a direct outburst of very strong winds. Storms of excessive strength with extreme wave height groupings may be attributable to composition of such winds with magnitudes of highly damped formation. Therefore thorough study and analysis of temperature and wind together may be essential.

16. Frequency distributions of observed wind direction should be presented in the form of wind rose with clear indication of the location of the station they belong, and with complementary comments of how different it would be at the specific jobsite in concern.

17. Wave: " HOWEVER BIG WAVES GET, THERE WILL ALWAYS BE A BIGGER ONE COMING," says so one theory of extremes, and experience suggests it is true. Wave height is the primary factor belonging to waves for the study of Rubble Mound Breakwaters, the factor most involved concerning the stability damage or run-up incurred on Breakwater, thus imprecision in measuring the wave height will result in scattering of the results. (Ref.4)

18. Frequency distributions of significant wave height to be provided in the form of a "Windrose" would be helpful for the correlation between wind direction.

19. Wave steepness and consequently wave period is a factor in the stability of Breakwaters. (Ref.5) gives criteria for steepest allowable slope, which could be a better specification to have such a control measure in supervision, by stating minimum slope and allowing no negative tolerance, while being prepared to pay for the excess naturally. It is known today that waves of more moderate heights could possess much greater steepness and asymmetry than the very large waves.

20. Construction needs to know wave criterias like water level fluctuations, cyclicity - periodicity, frequency of occurrance, energy of waves, direction, height, period of waves, that would be proving itself on the consequences of wave hydrodynamics; rather than those like; extreme wave heights, long term distributions, wave sequences, limited height of breaking waves, joint distribution of heights and periods, wave statistics, etc.

21. As most probably will be known, the weather shows a certain periodicity on a yearly basis. This is clearly demonstrated on Figure 1. In this respect it should be noted that wave heights are not the most important parameters. Limits in operation are mainly set by surging action that depends highly upon wave periods.

22. Not only the magnitude but the duration of storm resulting in water level build-up, thus changing application elevations of the maximum forces to much higher than those in design calculations or model tests, is essential to determine. (Ref.6)

23. Wave parameters which are mostly recorded are height and period. It is known that these wave parameters represent subjective observations based on certain discrepancies between visual observations and correspondingly measured parameters of significant wave height and zero crossing period, especially by extremely high and low waves. (Ref.2) Statistical distributions based on visual values with certain correction are, however, often used both for operational analysis and in order to determine the critical design criteria.

SOME PRINCIPLES

24. The construction team is always vitally interested in design data. To do his work properly he needs environmental data (e.g. wind, wave, humidity, temperature,), and to plan it he further needs information on model studies, soils, materials investigations, so that he will be aware what he may expect to encounter during construction.

25. Contractual obligations would draw construction people's interest to the definition of; " What is exceptional and what is not ?". However, his interest may well be concentrated over the frequency distribution of the governing storms with emphasis on wave height, period, frequency of occurrance - of other than maximum waves in the spectra, creating "Unworkable Conditions ".

26. Construction needs these informations on a limited scale. For his purposes of working conditions waves over 3 metres are not important, however, for the determination of the risks throughout the project naturally, he would wish to know all about extremes. (Ref.7) From Figure 2. one can see that, for H_s= 3 mt., p= 2.1% corresponding to 184 hours or merely 7 days in a year. If wave information could be supplied with adequate distribution (Periodicity-Cyclicity) throughout the year, having 2% probability of exceedance corresponding to 7 days/year, in my

opinion, construction team could satisfactorily prepare their planning of the project.

27. Construction people would wish to have data that will define the physical environment of the site rather than that of a place 500 miles away, where the information was collected from. Unfavorable conditions, emerging basically from wind/wave creating " Conditions Unsuitable for Working " depend on the details of a project, which could have affects over the operation at sea (Wave), over the operation of cranes with long booms (Wind), or of vessels with vulnerable wind surface.

28. Construction is in need of different data as well. In their interest of roughness the number of stormy days - adverse weather affecting his work - and its distribution in a year may be more valuable to him than the probability of exceedance of the significant wave heights. This could be supplied in a form of Figure 1., indicating the boundaries and the most probable pattern of distribution of the number of storm days in a month throughout the year, regardless of its direction. Additionally if the probability of exceedance beyond the upper boundary is provided, and if the definition of "Stormy Days" could be quantified relative to the method of construction - since 1 mt. or 3 mt. wave may be significant depending on the type of equipment - thus influencing the baseline of this pattern; it would be essential in the realistic planning of a project.

29. Field or prototype observations have not been given emphasis in the establishment of Design - Construction procedures, however, their soundness could be judged when compared to theoretical and probabilistic computations. (Ref.8 and Ref.2) Figure 3. shows that up to $H_s = 2.5$ to 3 mt. there is good agreement between data sources like wave measurements at site, ship recorded data for a wide region, naval oceanographic publications, and the Prototype Observations made; in spite of the fact that wave grouping of data would not provide detailed statistics for extreme events, and scatter in it would not allow extrapolation.

30. This may lead to the conclusion that Designer may always have adequate sources of information to provide the Construction group with such vital information, no matter how limited it may be for the design purposes, which would be quite sufficient for the purposes of Construction Planning, selection of method of construction and equipment, thus eventually allowing better costing of the project.

31. Observations should be made on Prototypes and if they prove to be safe and sound over a reasonable period only, then we can say that these parameters are adequate and are satisfactorily represented in the model. (Ref.9) ** NEC BABYLONIAS TEMPTARIS NUMEROS ! ** yes, do not trust random numbers and approach to the facts and benefit from the REAL DATA for the improvement of the set procedures of Design.

32. I do not understand how this field of ancient engineering, having roots from Phoenicians, have not kept pace with and benefitted from the miracles of modern electronics, simply by INSTRUMENTING THE BREAKWATER. I do believe there are lessons learnt from failures, but the Breakwater Prototype is still not observed and instrumented properly. The misfortune is to meet requirements in detail for air conditioned offices and cars for supervision crew in specifications, but not a bit of instrumentation for proper observation and capture of Real Data at the prototype.(Ref. 9) Sophisticated instrumentation for the capture and elaborate methods for the analysis of such data should be mobilized.

33. All of these things are not very difficult during the construction phase while the structure is in an analytical stage. There will not be problems of access to any corner of interest; huge equipment and manpower potential will be available; moreover lots of helpful details would exist, such as holes of tie-bars which could easily receive strain gages instead of being filled by epoxy mortar. Figure 4. shows such holes picturing water jets through them indicating pressure distribution.

34. While FAILURE MODELS against extreme conditions are necessary for better understanding of the prototype behaviour for better evaluation of costs; INTERMEDIARY MODELS with moderate conditions as well as exceptional would be essential for the testing of the intermediary - INCOMPLETE sections for better judgement of both the risks involved therein and the time scheduling of the construction period. Since these sections would be subjected to wave conditions different in both the frequency and the magnitude pattern than the extreme conditions against which they should also be tested to see the damage pattern of how the material would be dispursed, how could they economically be recovered, and the extent of RISKS that the contractor is subjected to. These incomplete - intermediary phases should be represented in model tests to verify soundness of designs for implementation.

35. Ability of safety related structure to withstand consequences of natural hazards depends on the damage creating conditions which not necessarily would be the one with probable maximum intensity. This is a long term planning, since projects - designs would be continuing in the meantime, with the understanding of the existing failures, lack of reliable data, and difficulties of representation in the model, it should at least be decided to define a level of risk and incorporate it into the design process.

36. Experience and judgement play a major role in the final outcome of a Breakwater design which poses broad and complex problems and is not readily subjected to detailed analysis. Even with the wave information, engineering judgement still needs to be exercised to take into account irregularities, or indeed, whether what was obtained was typical and over what period.

CONSTRUCTION REVISITED

37. "Would the lines carefully indicated in the DESIGN be realized exactly in the manner they look over the Drawings?" The answer is a definite NO, due to the difficulties met in their realization, or in other words "due to the difficulties met through the DESIGN phase for the proper representation of these realities in the DRAWINGS and SPECIFICATIONS!! "

38. The single directional observation of these discrepancies between the PROTOTYPE with that of its representation in the DESIGN raises the question which is the proper direction to look. Should the DESIGN be reflected in the PROTOTYPE or the PROTOTYPE be represented in the DESIGN ?

39. If " AS BUILT " drawings representing the PROTOTYPE finally built to the satisfaction and approval of the supervision, would present a completely different structure than that is sketched in the original " Approved working Drawings " of the DESIGN, how far are we going to insist that there is a satisfactorily reliable Design procedure ?

40. It is necessary to visualise the " Design and Build " phenomena as a continuous Decision making process throughout the stages of construction considering; availability of materials, equipment, manpower resources and their extent of use, impacts of environment, etc., leading to alterations or modifications in the set requirements of the design; thus enabling proper conduct and successful continuation of the construction as well as controlling the variations through it.

41. Wave information as height, period, frequency of occurrance and cyclicity of environmental conditions creating " Unworkable days ", is essential for the selection of " Floating Equipment " which has the advantage of shorter reach but disadvantage of movements during handling.

42. If, as usually it is at least for one season, floating equipment is going to be used during a season of " Less workable days ", where the probability of it being adversely affected by wave motion is high, the problem could partially be solved by planning its employment in a process like dumping - placing - some rubble material below the elevations susceptible to wave influence, much ahead of schedule. The most vulnerable technical difficulty is usually associated with placement of the core material which is left unprotected until protective sublayers are placed. It is therefore very important that the contractor schedules his work carefully during stormy seasons so that he by means of weather forcasting is able to seize advancement of core material or to protect it well in advance of adverse weather. However, the fact is even the method of construction could well be changed in view of such cyclicity information.

43. Not only the method of construction but also the type of equipment, the time scheduling, thus in consequence manpower, cost, overall planning functions are influenced by the degree of environmental information, available.

44. Failure can become a nightmare for the Engineer, whether he is engaged in design, construction, maintenance or operation. However, quite often the " Construction Team " bears the heaviest consequences of these " Facts of God".

45. It is obvious that a mound structure is not an easy structure to build, requiring knowledge and skills much beyond the rough mechanical experience levels. Experience in the construction of mound breakwaters has unfortunately often shown inadequate attention to the construction problems, as well as to the best possible use of the available resources, under the present local conditions. Therefore the Prototype will almost always be different from its Model, and sometimes to such an extent that the stability of the Prototype may have even been reduced. (Ref. 10)

46. Typically the decision maker not only does not know when a component would fail, but also he does not even know how reliable the component is in the sense of knowing the parameters which govern the failure process. Failures of mounds demonstrate many characteristic developments, like an S - shaped slope, however, analysis of the reasons for such damage were invariably not undertaken in detail at least by means of " Failure Models " simulating the prototype behaviour. The basic element of hydrodynamics could be better understood and the interaction between the hazard creating conditions and the structure could be judged only by observing the Prototype and definitely not in a model.

47. Construction people have a fair opinion upon the probabilistic nature of the design criterias which leads them to conclude their own assessment of the risks, emerging through the above mentioned uncertainties, which they naturally reflect in their construction cost estimates, to the allowed range of competition.

48. As at least one of the remedies of the subject deficiencies of the Hydraulic Models; further to the opinion speculated before that incomplete - interim phases of construction should be studied against not only the extremes but the modes of acting natural forces, Damage and Failure modes of Breakwaters should be investigated via " Failure Models " that would provide data for much better calculation of these Risks which inevitably would reflect in the devaluation of the UNKNOWN factors incorporated therein.

49. Defining the level of risk which can be accepted, and the availability of maintenance facilities for repair in the event of damage, and of course the cost of it, could all well be judged better with the provision of such " Failure Models ".

50. Generally, the economics of Breakwater construction and the difficulties of repair in the event of damage, as well as the loss of function it is serving for, favour a conservative approach to the determination of the size of critical elements like the main armour. However, I believe, if the consequences of the failure of a specific

part of Design could be simulated and evaluated by Interim and Failure Models, the positive reflection of it to the economy of the Project would be considerable.

CONCLUSION

51. Observance of structural or functional failures in Breakwater structures, in spite of the tendency to overdesign it, is the evidence that the existing Design philosophy must be changed.

52. To diminish the uncertainties - incompleteness in the data used for design, methodology and instrumentation is to be improved.

53. Complementary wide range of information, no matter how moderate it can be, foreseen by Construction group leading to proper evaluation of a project, which necessarily is not the same data of extremes sought by Designers, should be identified; and consistent definitions for the data to be specified should be developed.

54. The price of unreliability is very high, and the cure is in going to the basic by OBSERVING THE PROTOTYPE, evaluating its behaviour under the actions of real data. To establish a better Decision Making process and especially a more reliable Model Testing, Prototypes should be instrumented, observed, perceptions of the visual and real data should be reflected in the design procedures.

55. International semantics problems particularly over the results of hydraulic model studies pertaining to their damage and soundness criterias should be solved.

56. Hydraulic Model studies used for verification of the soundness of a design should be extended for the verification and assessment of the conditions of failure and/or damage of the structure for which considerable RISK can easily be attributable by means of " Interim and Failure Models".

57. Incorporation of level of Risk into the design process of such gigantic structures subject to huge environmental hazards is long overdue.

58. Construction experience and judgement should be included and method of construction in part-icular should be valued in the design process which should be treated as an ART and not solely as a scientific calculation.

59. Nomatter how profficient the methodologies and scientific computation tools are programmed into a technocrate, he still is not given emphatically the essence in developing the capability of using his JUDGEMENT , which in my opinion constitutes in real life applications, the most powerful talent that makes him become an ENGINEER .

REFERENCES

1. PLOEG J. On the importance of designing wave climates. The 6th International Conference of Port and Ocean Engineering under Arctic Conditions, POAC 31, 1981, Quebec, 809-819.
2. HALAND L. and SMALAND E. On the connection between observed and computed wave heights.
3. THOMPSON E.F. and HARRIS D.L. A wave climatology for U.S. Coastal Waters. Proceedings of Offshore Technology Conference, Paper No. OTC. 1693, 1972, Houston 675-688.
4. QUELLET Y. Considerations on factors in Breakwater model tests. Proceedings of 13th Coastal Engineering Conference, July 1972, Vancouver, 1809-25.
5. WALTON T.L. and WEGGEL R.J. Stability of Rubble Mound Breakwaters. Proceedings of ASCE journal of waterways and Harbours and Coastal Engineering, 1981, Vol.107. No. WW3, 195.
6. MESTA M.A. Design and Construction Interface International Symposium on Port and Ocean Engineering. March 1982, Mexico City.
7. GUNBAK A.R. Wave analysis for Tripoli Harbour. Appendix I of reference 8. below. December 1978, Fig. 15.
8. DANISH HYDRAULIC INSTITUTE (DHI). Report on Breakwater model tests. February 1979, 14.
9. MESTA M.A. Breakwater as Prototype. The 5th International Conference of Port and Ocean Engineering under Arctic Conditions, POAC 79, 1979, Trondheim, 1365-1383.
10. TØRUM A., MATHIESEN B., ESCUITA R. Scale and Model effects in Breakwater model tests. The 5th International Conference of Port and Ocean Engineering under Arctic Conditions, POAC 79, 1979, Trondheim, 1335-1350.

Discussion on Papers 1 and 2

N. HOGBEN (*National Maritime Institute Ltd*)
Professor Battjes begins by stressing the need to extract information from all potential sources to meet practical requirements for data. This is a view which I strongly endorse and it has indeed been a key theme of the work at the National Maritime Institute Ltd (NMI), to which he later refers in the very useful review of available information and analysis methods and which he then presents. Because it is often necessary in practice to derive wave statistics from indirect or inadequate sources it is very important that such data and the methods of derivation should be subjected to critical evaluation.

The work at NMI Ltd has been largely concerned with the development and validation of methods for enhancing the reliability of visual data which have the important advantage of world-wide availability but must be interpreted with great care. These methods are based on functional modelling of the statistical relations between wave heights, wave periods and wind speeds and are now incorporated in a computer program with the codename NMIMET. A crucial feature of this program is its capability for synthesizing reliable wave period statistics from visual wave height data without any use of visual observations of wave period which are known to be very unreliable. The basis of the analysis methods used in the NMIMET program is explained in the Paper, which cites some sample results indicating the reliability attainable in Figs 2 and 3 and quotes references to more detailed accounts.

The key point I wish to emphasize concerns the problem of the spatial variability of conditions. When using visual data it is in general important to choose a 'catchment' area which is large enough to yield an adequate observation count but not so large as to be unrepresentative of the conditions at the required location. In coastal areas this may be difficult to achieve because there will often be a high degree of spatial variability, and the observations will tend to be concentrated on shipping lanes where the conditions may be different from those at nearshore sites of interest to coastal engineers.

This point may be illustrated by consideration of the accompanying figures cited from ref. 1. Fig. 1 is a map showing 16 areas used in the validation studies for NMIMET including depth contours from which an impression of the likely spatial variability of conditions can be formed. Fig. 2 shows a comparison of measured and visual

wave height data for the area around Seven Stones and the close agreement indicates, as in the case of the results shown in Fig. 2 of the Paper, that conditions in the visual catchment area are representative of those at the measuring station. Fig. 3 by contrast shows a similar comparison for the area around the Shambles lightvessel. In this case the visual catchment area spans the main shipping lanes of the English Channel and, not surprisingly, the visual data are concentrated offshore and show a greater incidence of higher waves than those measured at the lightvessel which is in a sheltered nearshore position. In such cases, a need commonly encountered by coastal engineers for estimating nearshore wave statistics from offshore data arises, and I would welcome the Author's comments on how this can be most effectively done.

Refs 1 to 4 are cited to supersede refs 12 and 16 of Paper 1. Ref. 1, together with refs 14 and 15 of the Paper are detailed accounts of the work. Refs 2 and 4 describe various aspects of the work with emphasis respectively on validation, application and methodology.

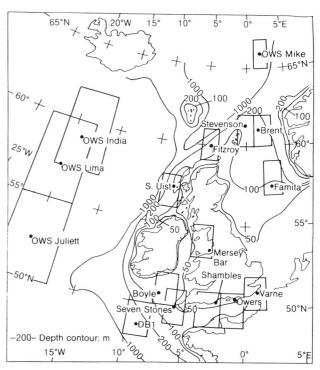

Fig. 1. Map to show 16 areas used for validating NMIMET

Breakwaters—design and construction. Thomas Telford Ltd, London, 1984

25

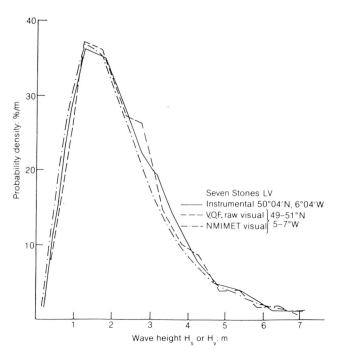

Fig. 2. Comparison of instrumental and NMIMET visual wave height probabilities for the Seven Stones area

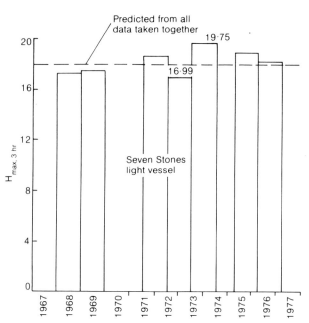

Fig. 4. Annual variations in wave height at the same site, return period 1 year (ref. 5)

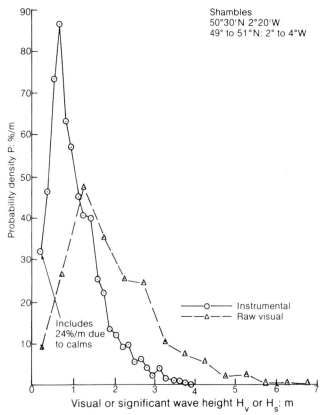

Fig. 3. Comparison of instrumental and raw visual wave height probabilities for the Shambles area

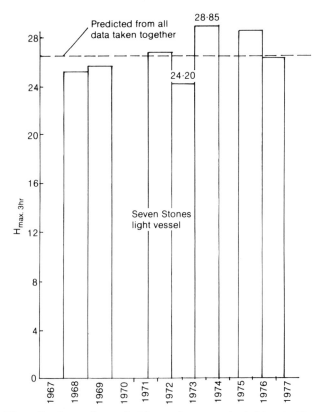

Fig. 5. Annual variations in wave height at the same site, return period 100 years (ref.5)

J. BERRY *(Bertlin & Partners)*
It is common to design breakwaters for the 50
year or 100 year wave. However, there is a 63%
probability of it occurring or being exceeded
during the 50 year or 100 year lifetime of the
structure.

Our wave climate data is inadequate as long
period recordings usually do not exist. Even
when (occasionally) they are available, there is
considerable variation between the individual
years. This is illustrated in Figs 4 and 5
which show graphically the results of 7 years
of wave records at Seven Stones lightvessel,
presented in a paper by Tucker & Fortnum of IOS
(ref.5). Values of $H_{max, 3h}$ for a return
period of 1 year (Fig.4) vary from a minimum of
16.99 m to a maximum of 19.75 m. When all the
data for the 7 years are combined and analysed
the value is 18 m. For a return period of 100
years, (Fig. 5) the predicted $H_{max, 3 h}$ values
vary from 24.20 m to 28.85 m. The combined
data give 26.3 m. The variation in predicted
wave height has an accentuated effect on the
weight of armour needed for a breakwater. This
is shown in Fig. 6 using the 100 year wave predictions illustrated in Fig. 5. Assuming a
cubic relationship between wave height and weight
of armour unit, the ratio between the weight
corresponding to the minimum 100 year prediction
and the maximum one is 1:1.69, a difference of
nearly 70%. Thus even when long-term records
are available there is a high degree of uncertainty as to their validity when estimating the
weight of armour required.

When designing building structures it is now
usual to use characteristic strengths of materials
and characteristic loads. Values are adopted
which have a 5% probability of being exceeded
during the lifetime of the structure.

I suggest that instead of adopting the once
in 100 year wave (or 200 year or more, if one
wishes to have an extra margin of safety), we
work towards the concept of a design wave
climate which has a low probability of being
exceeded during the lifetime of the breakwater.
I do not know whether this should be a 5%
probability or a 10% probability, but suggest
that the probabilistic approach is one which
designers should be pursuing.

W. F. BAIRD *(W. F. Baird & Associates, Ottawa)*
Professor Battjes questioned whether the wave
records (containing grouping) used in the
Canadian experiments were realistic. These wave
records were 20 min prototype records selected
because of the wave grouping that they displayed.

O. T. MAGOON *(US Army Corps of Engineers)*
I agree with Professor Battjes' point that
recent work in the USA supports the conclusion
that wave grouping in deep water may be accounted
for by conventional statistical procedures.

In shallow water, wave grouping may have
dramatic effects, e.g. jetty entrance, Nogo
Harbour, California.

W. A. PRICE *(Hydraulics Research Station Ltd)*
Professor Battjes made a point that short-
crested seas were perhaps more important for
gravity structures than for rubble-mound breakwaters. I feel that short-crested seas are

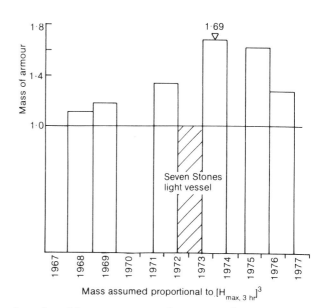

Fig. 6. Effect on mass of armour of annual
variations on wave height, return period 100 years

important for rubble mounds too. Paper 6 deals
with this subject.

H. F. BURCHARTH *(Aalborg University, Denmark)*
It holds for many breakwaters that overtopping
(and not only armour stability, etc.) is the
critical factor. It is to be expected that the
amount of overtopping and its distribution in
time and space must be greatly affected by the
lengths and directions of the wave crests.

P. LACEY *(Ove Arup & Partners)*
In a letter to the Dock and Harbour magazine,
published on 30 October, 1981, I appealed for
all the research, development and information
published in papers to be amalgamated to produce
a state of the art document. It is hoped that
this conference on breakwaters has progressed
a long way towards this.

Due to an inevitable search for economy our
studies have persuaded us to steepen the breakwater slopes to excess, aided by the apparent
properties of various man-made armour units.
The safety or assessment of failure risk,
indicated by the model tests, has not been
borne out by prototype breakwaters, which have
failed in both shallow and deep waters.

This factor, multiplied by the inability to
predict accurately environmental loadings,
leaves a burden on the design engineer. You
cannot get a 'gut engineering' feeling from a
model.

I would therefore like to pose a question
to all the authors that in the splitting of the
design process between research laboratories
and data acquisition and the design engineer,
have we not forgotten the principle used by our
famous predecessors, which was that the person
who designs the breakwater should supervise the
construction and the after-construction
monitoring.

L. DRAPER *(Marine Information & Advisory Service)*
The Marine Information and Advisory Service
(MIAS) has been established to locate and store
environmental design data and to act as a
consultancy, with a world-wide responsibility.

Questions are answered in-house or referred to the most suitable authority anywhere in the world. Data are exhaustively quality controlled before being accepted into the data bank; poor quality or uncertain data are rejected. At the behest of the Intergovernmental Oceanographic Commission of UNESCO, MIAS acts as the responsible National Oceanographic Data Centre for waves - a kind of World Wave Data Centre - for instrumental wave data. Thirty area representatives around the world are continually on the watch for such data, and now the World Meteorological Organization has asked us to participate.

P. LACEY *(Ove Arup & Partners)*
I believe Mr Mesta is gently tilting at consultants in his Paper and I would like to ask him:

(a) How does the contractor define risk during his construction?
(b) Should the contractor institute his own wave measurements and would this improve matters?
(c) Has Sezai Turkes-Feyzi Akkaya monitored its breakwater constructions? Has this changed the design process?

J. A. BATTJES *(Paper 1)*
Dr Hogben raises the question of how to estimate nearshore wave statistics from offshore data.

I would recommend the transformation of offshore data to the nearshore site using a numerical model, which should be calibrated if possible with local data. If no such data are available at the time of initiation of a project, a wave data gathering program can be included in the surveys. Even short series can be very valuable for purposes of calibration.

The numerical transformation referred to must generally include refraction, shoaling, and dissipation in the bottom boundary layer. In very shallow water (less than three to four times the local r.m.s. wave height) a dissipation by depth-controlled wave breaking must be included as well. The calculations must be performed for each of a discrete set of incident wave parameters.

The general principles underlying the above mentioned models are well known. I would now like to comment on some recent developments.

A common problem in the calculation of wave refraction in shelf seas and coastal areas is the chaotic ray pattern which results if the propagation distance is large compared to the length scale of the bottom topography, as it usually is. In Bouws and Battjes (ref. 6) the locations of calculated rays are treated as the results of random sampling. The variability is suppressed by an efficient spatial averaging procedure. In this viewpoint, only the spatial ray density in a certain area has meaning (the location of individual rays has no meaning). The method is efficient in the sense that calculation of individual ray separation factors (or refraction coefficients) is not necessary. It is used extensively in Dutch coastal engineering practice. A verification of the method for a case of periodic waves, including a caustic, is given in ref. 7.

A review of bottom-induced wave energy dissipation parameterizations has been given by Shemdin in ref. 20 of Paper 1. A recent contribution in this area has been given by Bouws and Komen (ref. 8), who were able to make estimates of the individual source terms in the spectral balance equation from measurements during steady-state conditions in a storm in the southern North Sea, where the restricted depth and the associated bottom-induced dissipation significantly lower the attainable wave energy levels compared with those in deeper water.

A model for the energy dissipation due to the depth-controlled breaking of random waves has been presented by Battjes and Janssen (ref. 9). The fraction of breaking waves is calculated probabilistically, and for those waves which do break the dissipation rate is estimated on the basis of an analogy with a bore. The theoretical results were found in good agreement with laboratory data (ref. 9), even on a barred beach with 'green water' in the trough between two breaker zones. An extensive comparison with field data is presently in progress. The results obtained so far show equally good agreement. The model is used increasingly in Dutch coastal engineering practice.

The preceding comments referred to individual transport terms and source terms in the wave energy balance equation. A related area of more recent developments concerns the spectral shape which emerges as the final results of all the contributory processes. It is shown in ref. 10 that the wind-sea spectrum in restricted depth has a self-similar shape which can be very well approximated as the product of a JONSWAP-shaped spectrum and a non-dimensional function which suppresses the spectral values for small values of the non-dimensional frequency $\omega(h/g)^{\frac{1}{2}}$. (This function was derived by Kitaigorodskii et al. (ref. 36 of Paper 1) for the saturation range only; see also paras 44 and 45 in the paper.) In ref. 11, a sequel to ref. 10, it is shown that many deep-water results concerning spectral parameters and their dependence on the growth stage apply also in restricted depth, provided that they are expressed in terms of wave number rather than frequency. (This empirical result is analogous to what Kitaigorodskii et al. had assumed about the spectral saturation level. It reconfirms that from a theoretical viewpoint wave number spectra are more fundamental entities than frequency spectra, because they have better invariance properties.) The results presented in refs 10 and 11 are being applied in a newly developed hybrid model for waves in water of restricted depth (see para. 46 of the Paper).

Mr Berry reminds us that there is a 63% chance of an event occurring in a time interval equal to its return period, and in the last paragraph of his discussion he suggests using a design climate with a lower probability of being exceeded during the lifetime of the breakwater. I agree with this suggestion, but I would not favour attempts to determine a priori a more or less universally applicable level of that probability. After all, the consequences of exceedence of the design conditions can be quite different between projects. I would therefore prefer a design procedure based on economic optimization, at least for major projects (see para. 48

of Paper 1, and Paper 10).

Mr Berry also gives some wave height data (Fig. 4 of his discussion) to illustrate that there is 'considerable variation between the individual years', which leads to 'a high degree of uncertainty' in estimates of the required armour unit weight.

Contrary to Mr Berry, I find the variation shown in his Fig. 4 not at all large. In fact, considering that the data are estimates of H_{max} in 3 hours with a return period of one year, rather than some annual mean value, I find their range surprisingly small.

Disregarding differences in subjective appraisal, the fact remains that there is a non-negligible variation. However, this can be dealt with using available statistical methods, such as those referred to in para. 57 of Paper 1. If it is Mr Berry's purpose to emphasize the need for long-duration data and for uncertainty assessment, then I wholeheartedly agree with him. In fact, these are two of the principal points I have been trying to make.

Mr Baird points out that prototype records were used in the Canadian experiments (on stability of breakwaters subjected to attack by wave trains with varying degrees of grouping). It is true that in the paper by Johnson et al. (ref. 12), results of stability tests are shown for simulated prototype records, but these authors state that tests were also run with 'an accentuated and rather artificially grouped wave train'. Apart from this subtlety, the fact remains that no systematic statistical investigation was reported in ref. 12 concerning the validity of the model of independent, uniformly distributed random phases for the prediction of groupiness in natural wind waves. My point was to emphasize the need for such investigation, and to refer to some results which have recently become available (refs 1 and 2 of Paper 1). These lend support to the random-phase model as far as the groupiness is concerned. Needless to say, the wave shape is greatly influenced by bound higher harmonics, but these do not appear to be signi-

ficant for the group formation. This was shown implicitly in ref. 1 and explicitly in ref. 2.

M. A. MESTA *(Paper 2)*
Mr Draper states that the main concern of MIAS is to collect and store instrumental wave data; the importance and value of these data are unquestionable.

However, I believe that these data are gathered for the definition of the environmental forces which, in Paper 2, I referred to as the 'data of action'.

The lack of data essential for the correct perception of the prototype behaviour, referred to in Paper 2 as the 'data of reaction', are the responses or reactions to be captured at the prototype structure itself, constituting of, but not to be limited to, Table 1 (ref. 9 of Paper 2).

Finally, reiterating the essence of better information on the 'data of action', I would like to quote Professor T. O'Riordan: 'Misperceptions are not easily removed, certainly not by the provision of more or clearer information!'

With regard to Mr Lacey's query, I was not tilting at consultants in the Paper - at least I did not mean to. In view of the importance of consultants, in their role in design, I intended to identify the mishaps in the practised procedures, in which we are all involved, with the hope of receiving their accord for the establishment of improvements.

To answer Mr Lacey's questions
(a) Every contractor interprets or accepts risk under different definition as well as magnitude. There is a long history of attempts to combine probability and consequence into a scalar measure of risk. The oldest and still most common way of combining probability and consequence is through their product, the expectation. However, at least since Bernouilli, there has been recognition that individuals do not value uncertain propositions by expectation. Low probability, high consequence risks are not interchangeable with those of high probability and low

Table 1. Data of reaction (ref. 9 of Paper 2)*

Instrument	To determine
Piezometers (hydraulic or pneumatic)	Pressure distribution at base of vertical wall and flow pattern and fluctuations throughout the rubble mound, permeability of each layer
Velocity meters (oedometer)	Uprush and downrush velocities
Current meters	Inflow - outflow velocities, directions, temperatures in the vicinity of structure
Pressure cells	Wave and dynamic shock pressures and distribution on slopes and vertical faces
Strain gauges, (or special paint with polariscope)	Horizontal and vertical stress distribution, leading to friction as well
Settlement gauges	Magnitude of movements, rate of consolidation-compaction in rubble mound
Tide gauges+	Water level fluctuations
Wave recorders+	Spectral info, wave height, period
Echo sounder or side scan+	Changes in the batymetry before and after construction
Laser tube	Deflection magnitudes, movements in walls

*Visualizing the hydrodynamics at and within the breakwater structure it would primarily be recommended to 'Instrument' the breakwater prototype with probably self-recording type gauges.
+Seldom applied

consequence ones. Typically, individuals willingly pay a premium to avoid high-variance risks, which are inherent in the nature of the construction aspects.

The Engineer must design a prototype to exist and survive in theory and practice, which makes the discussion of its reliability appropriate.

In the context of reliability, which is measured by the probability of a structure performing its purpose of design adequately for the period of time intended under the operating conditions encountered, I would define risks in the construction industry as those attributable to the reliability throughout all phases of construction including the liability period of the contractor.

However, if the interest is not in the assessment of risk by its full meaning in reliability engineering, then I would agree with Erikson (1978) in the definition as 'Risk is an exposure to economic loss or gain arising from involvement in the construction process'.

(b) Yes, the contractor should institute his own wave measurements, which is usually done, but this would only be effectively beneficial during the construction period. However, the wave information referred to in Paper 2 should be furnished to the contractor before the initiation of the contract, namely in the tender period, for it to have proper evaluation and thus positive reflections over the assessment of the project.

(c) I personally, and also S.T.-F.A. as a company have closely monitored their breakwater constructions. I have stressed the importance of prototype monitoring in a paper to POAC '79 (ref. 9 of Paper 2).

My monitoring of the prototypes is reflected in the design processes applied to numerous 'turn-key' and 'design and build' contracts performed in the international field. This, I believe, constituted the major reason underlying such success in competitiveness.

REFERENCES

1. ANDREWS, K.S., DACUNHA, N.M.C. and HOGBEN, N. Wave Climate Synthesis. NMI Report R149, January 1983.

2. ANDREWS, K.S., DACUNHA, N.M.C. and HOGBEN, N. Assessment of a New Global Capability for Wave Climate Synthesis. Oceans '83, San Francisco, August 1983.

3. ANDREWS, K.S., DACUNHA, N.M.C. and HOGBEN, N. Ocean Wave Statistics: A New Look. Oceanology International Conference, Brighton, March 1984.

4. ANDREWS, K.S., DACUNHA, N.M.C. and HOGBEN, N. Wave Climate Synthesis Worldwide. Wave and Wind Climate Worldwide, Roy.Instn Nav.Archit. London, April 1984.

5. TUCKER, M.J. and FORTNUM, B.C.H. The significance of the N-year design wave height. In The management of oceanic resources - the way ahead. Vol. 2, pp.207-224. Proceedings of International Conference on the management of oceanic resources, London, April 1981. Los Angeles: University of Southern California, Institute for Marine and Coastal Studies, for ECOR.

6. BOUWS, E. and BATTJES, J.A. A Monte Carlo approach to the computation of refraction of water waves. J. of Geoph. Res., 87, C8, 1982, 5718-5722.

7. BERKHOFF, J.C.W., BOOIJ, N. and RADDER, A.C. Verification of numerical wave propagation models for simple harmonic linear water waves. Coastal Eng., 6, 255-279, 1982.

8. BOUWS, E. and KOMEN, G.J. On the balance between growth and dissipation in an extreme, depth-limited wind-sea in the southern North Sea, March 1983.

9. BATTJES, J.A. and JANSSEN, J.P.F.M. Energy loss and set-up due to breaking of random waves. Proc. 16th Int.Conf.Coastal Eng., Hamburg, I, 569-587, 1978.

10. BOUWS, E., GÜNTHER, H., ROSENTHAL, W. and VINCENT, C.L. Similarity of the wind wave spectrum in finite depth water. Part I - Spectral Form. April 1983.

11. BOUWS, E., GÜNTHER, H., ROSENTHAL, W. and VINCENT, C.L. Similarity of the wind wave spectrum in finite depth water. Part II - Statistical relationships between finite-depth spectral parameters and growth stage parameters, 1983.

12. JOHNSON, R.R., MANSARD, E.P.D. and PLOEG, J. Effects of wave grouping on breakwater stability. Proc. 16th Int.Conf. Coastal Eng., Hamburg, III, 2228-2243, 1978.

3 Durability of rock in breakwaters

A. B. POOLE, DPhil, BSc, FGS, P. G. FOOKES, DSc(Eng), PhD, BSc, FIMM, FGS, T. E. DIBB, PhD, MSc and D. W. HUGHES, BSc, Queen Mary College

Field and laboratory studies carried out for the past two years in UK, the Middle East and Australasia are summarised. This work is continuing but provisional conclusions are given, together with recommendations on test specifications for rock.

INTRODUCTION

1. Rock has been used in harbour works and shore protection for over 5000 years and rubble mound breakwater structures have been armoured with rock for well over 100 years. In the past three decades the use of rock has increased significantly in many parts of the world and its durability in the marine environment has become a matter for concern.

2. The mechanisms which lead to the gradual deterioration of rock armour blocks on breakwaters are essentially physical in character. Chemical degradation processes also occur but are generally of minor importance, they include the gradual solution of carbonate rocks, the solution of salts and the oxidation and hydration of iron compounds such as oxides and sulphides which may cause spalling of surface material through volumetric expansion on alteration.

3. The physical processes causing the deterioration of rock and concrete armour units on breakwaters were discussed by Fookes & Poole in 1981 (ref.1); this paper outlines results of laboratory and field studies subsequently. In detail the mechanisms operating on rock armour are complex and interrelated but it is possible to group them under three main headings, Dibb et.al. (ref. 2), viz: abrasion mechanisms, surface spalling and major or 'catastrophic' fracture. These mechanisms may all be regarded as disaggregation of the rock fabric with the principal difference being one of scale. Examples of the results of the three types of mechanism are illustrated by Figs. 1, 2 and 3.

4. **Abrasion** is the term used here for any process which essentially removes the individual particles of the surface layers of rock by external action. This includes attrition by sand in suspension, rocking of one armour block against another and even the effect of the sea alone on soft or weak material which can be washed out of cavities or joints by hydraulic pressures.

5. **Spalling of surface** layers of rock can be achieved by a number of processes, salt attack, i.e. sea salts crystallising out in rock pores,

is perhaps commonest, but freeze/thaw thermal movements, alteration of minerals or expansion of clay minerals, can all lead to surface spalling.

Fig. 1. Examples of armour blocks rounded by abrasion.

Fig. 2. Spalling of the surface layer of a limestone armour block due to salt attack.

6. **Catastrophic fracturing** in this context refers to the splitting of the rock blocks into two or more major parts. Typically such fractures follow incipient planes of weakness in the rock which may be due to incorrect handling procedures in the quarry, or in transportation or placing. Alternatively they may arise directly as the result of block movement on the breakwater during storm conditions.

Breakwaters—design and construction. Thomas Telford Ltd, London, 1983

31

Fig. 3. Major or 'catastrophic' fracture across a block of primary rock armour.

7. It is clear that all of these processes can act together or singly and also that there will be overlap between them with one type of deterioration assisting and enhancing another. It is, therefore, often difficult to evaluate the relative importance of the different mechanisms on a particular structure since usually only the final result of the degradation can be seen.

FACTORS CONTROLLING DETERIORATION

8. The relative importance of the various deterioration mechanisms are controlled by the factors
   a) Position on the breakwater
   b) The local meteorological climate
   c) The local physical environment
   d) The rock type and its weathering grade.

9. The position of the armour block on the breakwater face is of obvious significance. This is recognised in Fig. 1 of Fookes & Poole (ref.1) which identifies four horizontal zones on the face of the breakwater, each with a characteristic pattern of decay mechanisms operating. In practice it has been found that field study observations do not always clearly differentiate all the four zones, and the simplified threefold

zonation given in Fig. 4, is perhaps more appropriate to field studies of armour block deterioration, interlock and assessments of breakwater damage. This simplification is justified when abrasive rounding of blocks is considered. Dibb et.al. (ref.3) have shown that two breakwaters of differing ages from different environments have similar patterns of rounding with a single maxima in the intertidal zone although measured on a four zone basis. This pattern is indicated in Fig. 4.

10. The meteorological climate is significant in that it will change the order of importance of the degradation processes operating. The degradation arising through physical salt weathering - salt attack, is found to be most important in hot desert environments where evaporation leading to salt crystallisation is at its most effective and where salt accumulation on surfaces is possible without being flushed back into the sea by rain. Salt is perhaps most effective as a decay mechanism when it accumulates in the fine grained dust that is found in the supra-tidal zone between armour blocks. This dust behaves as a salt pan, occasionally wetted by sea-spray and condensation. Degradation of rock armour in such an environment can be rapid and serious. By contrast salt attack in regions where evaporation is low or rainfall high is of much less importance although disruption of rock surfaces due to crystallisation in cracks can occur under most climatic conditions. Freeze-thaw processes in cold climates is in many ways an analogous mechanism in that the expansion of water on freezing, wedges open and expands the cracks in which it occurs. Hudec (ref.4) for example, has suggested that on a microscale simple freeze-thaw mechanism is inappropriate and water molecules are absorbed on to micro-crack surfaces and initiate the expansion and disruption.

11. The abrasive processes and catastrophic fracturing of blocks are little affected by meteorological climatic considerations and may be

Fig. 4. The environment zones of a marine breakwater.

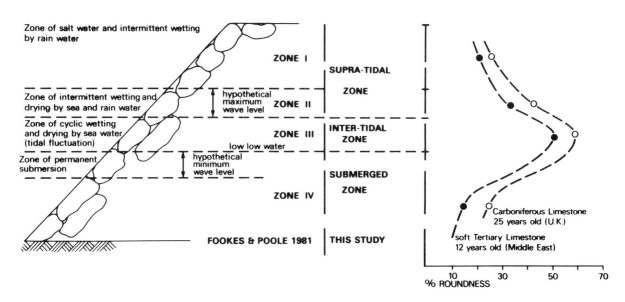

better correlated with position on the breakwater, the local physical environment or the rock types.

12. The local physical environment such as the depth of water, wave climate and storm frequency are principal design considerations for a breakwater structure, e.g. Darbyshire and Draper (ref 5). The detailed wave climate information is also of importance in assessing the relative severity of the various decay mechanisms operating. The breakwater structures along the east coast of Australia may be taken as an example since part of that coastline is protected by the Great Barrier Reef. Thus the structures along the New South Wales and Southern Queensland coast are most vulnerable to the high energy wave climate of the Pacific Ocean and typically have 'catastrophic' fracture of armour blocks as an important type of failure. Further north the Barrier Reef takes much of the energy out of the waves and the problems of catastrophic failure is reduced in importance.

13. In all the breakwater structures studied, it is evident that geographic and local physical conditions have a modifying influence upon the relative importance of the decay mechanisms. If

Table 1. Types of damage to primary limestone rock armour on three breakwaters in different environments expressed as a percentage of the total damage.

		1. Middle East	2. U.K.	3. E.Australia
Supra-Tidal Zone	Cavities	25	27	50
	Fractured Blocks	20	24	17
	Sub-Size Blocks	10	21	33
	Unstable Blocks	45	28	0
Inter-Tidal Zone	Cavities	18	32	69
	Fractured Blocks	32	36	16
	Sub-Size Blocks	14	5	5
	Unstable Blocks	36	27	10

1. Middle East: Hot/arid with low energy wave climate.
2. U.K.: Temperate/wet with moderate energy wave climate.
3. Australia: Sub-tropical with high energy wave climate.

damage is assessed in the field and classified according to type over large sections of breakwater face, breakwaters of similar type but in different environments, as is shown in Table 1, give some indication of environmental influence on the damage type.

14. Assuming that the rubble mound breakwater structure has been well designed perhaps the

most important single factor controlling the in-service durability of the rock armour units is the rock type itself and its grade of geological weathering. It is possible to estimate the grade of geological weathering, as found in the quarried rock, in a general way in terms of its petrographic features and engineering test results, see Fookes et.al. (ref.6) and BS 5930 (ref.7). With the few exceptions noted previously (para.2) little additional weathering of the mineral constituents takes place in engineering time scales. However, as rock strength is a function of the rock fabric and the natural flaws which occur in it, weathered rock with its microfractures and poor adhesion between mineral grains is noticeably less durable in the breakwater environment than fresh rock of the same type.

15. The discontinuity spacing (bedding planes, joints etc.) of the rock in the quarry is of primary importance since they largely control the maximum size of armour blocks that can be produced. Methods of maximising the output of primary armour stone blocks by appropriate blasting techniques are summarised by Dibb et.al. (ref. 3) but incipient flaws in blocks may result in fracture during transport or placing, particularly if stockpiling with consequent additional handling steps is involved. Clearly catastrophic fracture along discontinuities reduces the output of primary sized armour blocks from the quarry, but minor flaws and discontinuities in the fabric of the rock are also important in that they assist the subsequent abrasion and spalling processes occurring in service on the breakwater.

MECHANISMS OF DETERIORATION AND THEIR EFFECTS
16. The rounding of rock armour blocks on a breakwater face due to abrasive processes is an important consideration for the design and construction of the structure. As blocks become rounded during the life of the structure, their weight is reduced, the block to void ratio of the armour layer changes and the interlock between an individual block and its neighbours is reduced. The weight and interlock changes tend to reduce the stability of the armour layer while changes in void ratio alter the wave energy dissipating characteristics.

17. In order to quantify these effects either in model studies using wave flumes or by analysing damage patterns that develop on actual breakwaters, it is first necessary to quantify the observed rounding of blocks on the breakwater. Many methods of defining the shape and roundness of particles have been proposed and are published in the literature though the great majority of methods are applied to particles of sand grade size and smaller. Further limitations are imposed by the breakwater environment as the three dimensional aspects of the blocks are difficult to examine and measure. Study is further restricted as normally only the supra-tidal and intertidal zones are easily available for inspection.

18. Although the initial studies presented here were made by direct manual measuring techniques, much of the data was compiled from greatly enlarged photographs of sections of breakwater face. These photographs typically covered areas of between 50 and 300 sq.m. (depending on block size) so that at least one hundred blocks were available for direct measurement on the photograph. Although the data was essentially two dimensional, the simple method for determining percentage roundness proposed by Krumbein (ref.8) and Fookes & Poole (ref.1) has proved sufficiently precise and reproducible. In summary this method compares the average radius of curvature of the corners of the block in silhouette, against the diameter of the maximum inscribed circle expressed as a percentage. The method remains somewhat tedious to apply manually and current investigation is being directed towards the automation of data collection and handling.

19. Examples of rounding of rock armour of various sizes and rock types have been measured on more than twenty breakwaters in the United Kingdom, the Middle East and eastern Australia. Rock type, block shape, weight and age of the structure vary considerably. If a given rock type, on similar structures in similar environments but of different ages, is examined, the differences in roundness can be easily observed in a qualitative manner although the differences in roundness value are quite small. Loss in weight due to rounding is not easily assessed by field or photographic measurement unless the blocks used are very regular in shape. However, comparison of block roundness measured on the breakwater, with samples of the same rock type artificially rounded in the laboratory, does allow some estimations concerning weight loss to be made.

20. A number of breakwaters on which block roundness percentages were calculated are tabulated in Table 2 with rock type, armour weight, block shape and age of the armour layer. The influence of rock type and age of the structure are clearly reflected in the roundness values. It is also clear from the data that rounding is much less severe in the supra-tidal zones when compared with the inter-tidal zone as is indicated in Fig. 4. In general terms limestones are beginning to emerge as the rocks most prone to rounding.

21. Measurement of block to void ratios for armour layers was attempted by various techniques and three were considered in detail and compared using data from four different breakwaters. A simple line count method was found to be the most satisfactory and reproducible of the three techniques.

22. Void ratios for a series of rock armoured breakwaters are given in Table 3. There appears to be little correlation between age of the structure and void ratio, even when similar rock types, such as those of the Middle Eastern examples, are compared little correlation can be found. This suggests that block shape and

Table 2. Percentage roundness of primary rock armour blocks in service.

Intertidal Zone

Age in years[+]	Rock Type	Armour Weight (Tonnes)	Armour Roundness %	Block Shape
U.K. Examples				
100)	Granite	.3	20-40*	Equant
to)	Limestone	.3	Up to 70	Prolate
1)	Sandstone	.5	25-30*	Tabular
17	Limestone	12	20-25*	Equant
17	Sandstone	12	20	Tabular
82	Pre Cambrian Melange material	5	19	Equant
	Slate	3	17	Tabular
	Grit	2.5	22	Irregular
Middle East Examples				
12	Limestone	8	61	Equant Irregular
14	Limestone	8	63	Irregular
8	Limestone	8	57	Irregular
4	Limestone	6	48	Irregular
E. Australian Examples				
1	Granite	5-12	17	Equant
13	Granite	5-8	26	Equant
15	Tonalite	Up to 10	28	Tablate Equant
8	Limestone	6-15 max	34	Equant
17	Basalt	10-15	37	Equant
28	Diorite	15	22	Irregular

Supratidal Zone

UK and E. Australian Examples				
100)	Granite	3	15-25*	Equant
to)	Limestone	3	Up to 45	Prolate
1)	Sandstone	5	15-30*	Tabular
8	Limestone	6-15	29	Equant

* Range of values from different sections of the same breakwater.
+ Length of time that groups of armour blocks have been in place on the breakwater.

Table 3. Void ratios of primary rock armour layers in service.

Armour Weight (Tonnes)	Void Ratio %	Length of Line (m)	Age of Structure. Years
1 - 3	36	102	up to 100
5 - 20	29.5	85	2
2 - 12	32	60	3
2 - 5	28	100	4
2 - 8	32 - 41	96	12
2 - 8	34	100	8
2 - 8	29	120	14
2 - 10	36	100	7
Up to 20	31	80	25 approx.
10 - 15	27	80	17
Up to 10	32	80	15
6	28	160	8
Up to 12	38	80	28
5 - 12	33	100	90

placing have a greater influence on void ratios than subsequent rounding of the blocks in service.

23. Spalling of rock surface layers as the result of thermal movement, alteration of iron oxides and sulphides or salt attack can be important mechanisms for the degradation of rock in a marine environment. Sperling & Cook (ref. 9) review the theoretical and observational aspects of salt induced deterioration of rock materials. They note that certain salts are clearly more destructive than others and account for this as being the result of the pressures exerted by the salt during its hydration. Thus sodium sulphate (with its septahydrate and decahydrate forms) is more destructive in the soundness test ASTM C88 (ref. 10) than sodium chloride (with no hydrate form) used in a similar manner. Chapman (ref. 11) considers that crystallisation rather than thermal expansion due to diurnal heating of salts is the main cause of pressures where halite (sodium chloride - no hydrate form) alone is concerned, the effectiveness of such pressures is illustrated by the electron micrograph (fig. 5).

Fig. 5. Scanning electron micrograph of halite (NaCl) crystals growing in a microcrack in dolerite. Width of field of view 600μm.

24. The effect of spalling of rock surfaces is to accelerate the rounding processes by abrasion of armour units and therefore in field investigation at a statistical level, and in measurements taken from field photographs rounding due to spalling cannot be differentiated from other rounding processes. It may be expected that spalling due to salt attack would be one of the causes for the rapidity with which limestone blocks round in the Middle East environment as may be seen from Table 2. However, it must be noted that the porosity of these soft limestones greatly facilitates salt attack.

25. Catastrophic fracture of blocks leads to the development of sub-size material in the primary armour layer which may be washed out during storm conditions to leave vacancies or cavities, or alternatively may infill existing voids so reducing the macro porosity of the armour layer. Since each fracture of this type produces two or more sub-size blocks the best approach to assessing the relative importance of this mode of failure, is to carry out damage assessments of selected sections of the armoured breakwater face. It has not proved possible to make assessments by photographic means but detailed field survey allows damage to be assessed under the following headings -

a) Cavities - gaps in the primary armour layer.
b) Fractures - armour blocks broken into two or more major pieces but still in place.
c) Sub-standard - below specification in weight or quality.
d) Unstable - blocks subject to movement in normal sea conditions.

Thus catastrophic fracture assessments may be related to items b) and c) in the above list. Examples of this type of damage assessment for a number of breakwater structures is given in Table 4. A crude assessment of the total damage to a given structure is shown in the last column of the table and is obtained by allotting equal weighting to each type of damage and expressing the total damage summation as a percentage.

Table 4. Damage assessments of sections of primary rock armour on structures in the UK, Australia and the Middle East.

Age Years	Armour Weight Tonnes	No. Blocks Studied	Cavities	Fractured	Sub-Standard	Unstable	% Damage
Middle East Examples							
8	8-15*	86	5	4	2	9	23
14	8	136	0	3	0	13	12
14	15*	300	0	7	1	2	0.3
12	8	60	8	1	0	9	30
12	8	62	10	0	0	4	22
2-3		150	0	0	0	0	0
4	1-6) 15*)	288	10	6	0	40	20
E. Australian Examples							
>5	5	542	28	1	26	8	11
17	10-15	131	11	6	0	1	14
15	10	434	23	7	8	16	11
15	10	72	6	2	6	0	19
15	10	71	14	3	1	2	28
8	6 6*	130	13	2	14	1	24
8	6 6*	146	11	7	9	1	20
28	10	113	67	3	0	11	41
1	5-12	73	13	0	2	3	24
1	5-12	72	5	1	6	3	21
1	5-12	96	6	4	7	3	21
1	5-12	201	24	5	15	6	25
12	8	261	22	6	19	15	23
U.K. Examples							
17	12,10*	257	----------------------				13
17	12,10*	247	----------------------				6

* denotes concrete armour unit

26. The progress of deterioration of the primary armour layer in the breakwater may be considered to take place in three stages:
1. Development of fractured, unstable and sub-size blocks as a result of the various deterioration mechanisms.
2. Development of cavities in the primary armour layer.
3. Destruction of the secondary armour layer and collapse of the structure.

The progression from stage 1 through to 3 usually does not take place in normal conditions and with the exception of unstable blocks and the gradual rounding of blocks by abrasion, deterioration is slow. However, storm conditions can cause a rapid progression to stages 2 and 3. Additional damage assessment studies currently in progress should allow a more detailed analysis of damage types, damage progression and rates of progression in a series of break-water environments. One conclusion drawn from Table 4 is that cavities are more common in the damage assessments of the eastern Australian examples compared to those of the Middle East in the low energy wave climates of the Arabian Gulf. Other forms of damage, particularly unstable blocks, are more common on the Gulf structures.

27. Although the percentage damage quoted for the various breakwaters is a simplification of a complex situation, the visual differences in total damage are easily seen. Figs. 6, 7, 8 and 9 are photographs of sections of breakwater faces with damage assessments from 0 up to over 30%

Fig. 6. Section of a primary rock armour layer 0-10% damage.

Fig. 7. Section of primary rock armour layer 10-20% damage.

Fig. 8. Section of primary rock armour layer 20-30% damage.

Fig. 9. Section of primary rock armour layer > 30% damage.

RATES OF DETERIORATION

28. Gathering sufficient data from a significant number of breakwaters to correlate rate of armour deterioration with rock type, environment and meteorological climate is a difficult task. However, a sufficiently large number of break-water structures with limestone and granite armour covering a range of ages have been examined and to allow the curves, Fig. 10 to show how rounding of armour blocks of these rocks progresses with time. The breakwaters selected to

Fig. 10. Progressive rounding of Australian granite and Middle East limestone armour blocks with time.

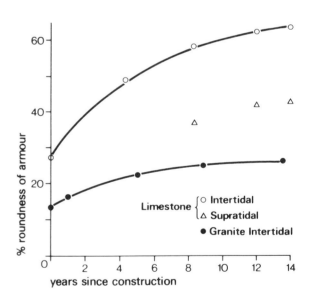

provide the data for these studies were sufficiently similar to allow comparisons to be drawn, though the environmental conditions for the granite armouring were more severe than for the breakwaters with limestone armouring. The form of the curves obtained was interesting in that rounding proceeds rapidly at first and gradually reaches a limiting roundness for the blocks after which weight loss through removal of material will continue though the block roundness value remains constant.

29. In view of the difficulty of obtaining roundness against age data from the field studies, alternative methods of assessing the rate by which abrasive rounding takes place, using laboratory tests, were considered. Certain standard tests, for example, the Los Angeles Abrasion test ASTM C535 and C131 (ref. 12), provide a measure of abrasion resistance but a closer comparison to the types of rounding processes operating in the breakwater environment, was obtained using a specially designed roller mill with a polypropylene drum.   In this new test marked blocks of the rock under test were tumbled together for specific periods of time.   The blocks were removed from the mill at set intervals and photographed in silhouette so that their average roundness could be determined using Krumbein's method (ref. 8).

30. Three British carbonate rocks were studied in this way, chalk as an example of the least durable materials likely to be encountered, a Jurassic limestone and a Carboniferous limestone. The pattern of their rounding with time is shown by the three curves in Fig. 11.   Weight loss with time in the roller mill was also determined as is illustrated for Jurassic limestone in Fig. 12.   In order to plot the curves shown in Fig. 11 roundness values were determined on a number of blocks of each rock type after time intervals in the roller mill from 0 to 50 hours.   In total 22 sets of measurements were made (Dibb et.al. ref. 3).

Fig. 11. Progressive rounding of 200-300g. blocks of British chalk, Jurassic and Carboniferous limestones in a 250mm diameter roller mill rotated at 20 rpm.

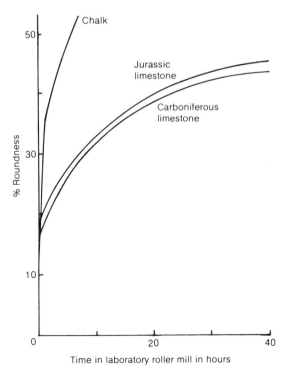

Fig. 12. Progressive loss in weight of British Jurassic limestone blocks with time of abrasion in the roller mill.

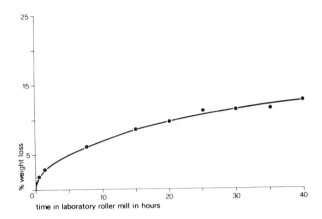

31. The comparison of the roundness values obtained from field measurements illustrated in Fig. 10 and the laboratory rounding curves for the limestone suggests that the laboratory test method is a satisfactory model of the rounding which takes place in the breakwater environment.   Although the data available is limited tentative correlations between field and laboratory results can be drawn for carbonate rocks.   Field data for Middle East breakwaters with a soft limestone armour indicates 50% rounding is reached after 5 years service (intertidal zone) while similar rounding is achieved in the laboratory after 1 hour in the roller mill.   The Port Talbot · breakwater with a Carboniferous limestone armour layer has reached 30% roundness after 20 years service (intertidal zone).   To reach a similar roundness in the roller mill Carboniferous limestone blocks require 15 hours in the mill.   The Carboniferous limestone on the Dawlish Warren sea defence works have a roundness value of 70% for the oldest material - estimated to be between 80 and 100 years old.   An equivalent rounding in the laboratory roller mill can be estimated by extrapolation of the curve as about 80 hours of milling.   For chalky limestone one hour of milling is equivalent to 5 years of rounding on the breakwaters in the Middle East, while with the stronger Carboniferous limestones one hour in the roller mill is equivalent to 1.2-1.3 years of service on UK breakwaters.

ROCK DURABILITY AND ENGINEERING TESTS.
32. A wide variety of standard tests are available for the assessment of the quality of rocks to be used as aggregates and·building stones.   The mechanisms of abrasion, spalling and fracture, noted as the principal mechanisms of deterioration operating on breakwater armour stone, are essentially mechanisms of disaggregation, thus discontinuities, flaws and rock fabric are perhaps the most important factors to be considered in testing the durability of rock for breakwaters. The range of tests that might be considered appropriate are -

a) Fracture toughness (K_{1c}).   Notched beam method ASTM E399-78a (ref. 13).

b) Aggregate impact value BS 812 (ref. 14).
c) Franklin point load test, Broch & Franklin (ref. 15).
d) Water absorption and relative apparent-density BS 812 (ref. 15).
e) Sulphate soundness ASTM C88 (ref. 10).

33. Wakeling (ref. 16) included the 10% fines value BS 812 (ref. 14) in his list of appropriate tests. However, because of the nature of deterioration mechanisms and the difficulty of avoiding some element of shear failure during test, compressive strength is not considered a satisfactory test in this context. Fracture toughness is considered an important if not fundamental parameter for rock strength since it will take account of the naturally flawed nature of the material. A detailed discussion of the aplication of fracture toughness tests to rock materials and the correlation of K_{1c} with other mechanical tests is given in Dibb et.al. (ref. 2). The relation between fracture toughness and the tests listed above is good as may be seen from Figs. 13 and 14. The data from which these figures are plotted is given in Table 5.

34. The soundness test used in this study was the modified version as described by Hosking & Tubey (ref. 17). It should be noted, however, that sulphate soundness test results cannot be directly correlated with salt attack in a marine environment because the temperature cycles used for the test are more extreme than would be encountered on the breakwater. Also the mechanisms producing the deterioration of the rock

Fig. 13. The correlation between fracture toughness (K_{1c}) and aggregate impact value for 10 rock types (see also table 5).

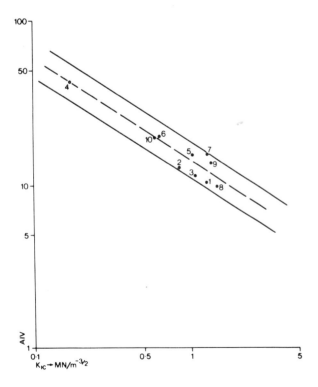

Fig. 14. The correlation between fracture toughness (K_{1c}) and magnesium sulphate soundness test results for 10 rock types (see also Table 5).

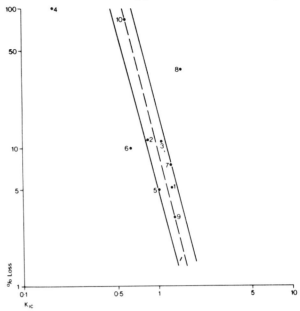

Table 5. Physical test results for various rock types.

Rock type	K_{1c} MN/m$^{-3/2}$	Soundness loss %	AIV	Key
Fine grained Carboniferous limestone	1.24 ± 0.17	5.2	10.5	1
Crinoidal Carboniferous limestone	0.825 ± 0.06	11.5	12.9	2
Chalk	0.170 ± 0.05	100.0	42.7	4
Dolomite	0.01 ± 0.02	5.0	15.3	5
Arkose	0.623 ± 0.06	10.0	20.0	6
Quartzite	1.23 ± 0.08	7.6	15.4	7
Dolerite (Weathering grades II to >III)	1.44 ± 0.31	36.5	9.9	8
Granite (Weathering grade I)	1.31 ± 0.11	3.2	13.8	9
Amygdaloidal Basalt (Weathering grade II to >III)	0.568 ± 0.10	82.8	19.5	10
Jurassic limestone	1.04 ± 0.02	11.1	11.5	3

with magnesium sulphate used in the test, are different from the mechanism appropriate to sodium chloride from sea water. Although the test is a useful one, interpretation of results requires care and the test cannot be simply regarded as an accelerated version of naturally occurring salt attack.

35. A more thorough treatment of the interrelationships between fracture toughness values and other engineering tests for a wide variety of rocks is currently in preparation.

36. The determination of progressive rounding of blocks abraded in the roller mill experiments are very time consuming to measure and as already noted only three carbonate rocks have been studied in this way. Other rock materials and correlations with standard abrasion tests are in progress and will be published in due course. If, as has already been suggested, the roller mill rounding results do reasonably represent the rounding processes which take place on the breakwater, then the standard engineering tests listed above should correlate with the rounding results obtained if they are appropriate tests for rock durability in the breakwater environment. Although the data currently available is limited to three carbonate rock types, plotting of roundness after a particular time interval in the roller mill against the results of the various engineering tests, gives a good correlation whatever time interval in the mill is chosen. To illustrate this the family of curves obtained for each test value after 2.5 hours rounding in the roller mill are plotted on Fig. 15. These results are encouraging and suggest that these engineering tests, with some additional studies, could be used to provide basic information concerning the relative resistance to rounding abrasion in the breakwater environment.

TEST METHODS AND TEST CRITERIA
37. If the laboratory roller mill experiments are an appropriate method of estimating the effect of

Fig. 15. The correlation of engineering tests with percentage roundness results.   After 2.5 hours of laboratory rounding.

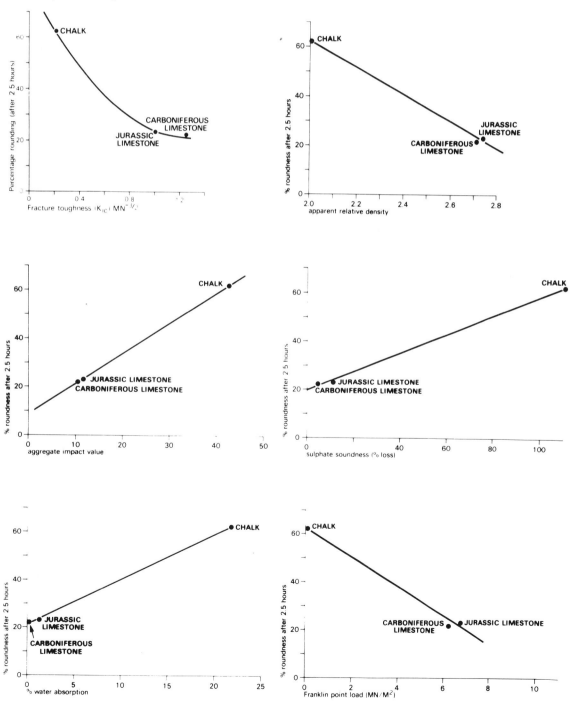

the rounding processes operating on armour blocks, then it should be possible to predict the behaviour of rock armour on the breakwater during its service life. Hence, it could allow the design of the structure to take account of the modifications to shape and weight that would occur during the planned life of the structure. It is already possible as a result of the study so far to draw a series of curves showing the rate of rounding in the laboratory roller mill (Fig. 16). Hours in the mill can also be related to rounding in service on the breakwater. Therefore, zones of acceptable and unacceptable material can be delimited (as illustrated on the basis of its laboratory rounding characteristics).

Fig. 16. Percentage roundness against time in the roller mill and suitability of carbonate rocks for primary armour.

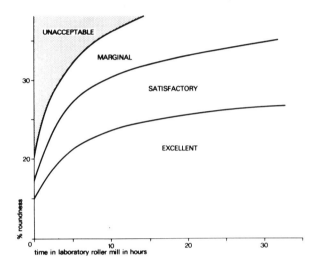

38. As has been suggested (para. 32), fracture toughness is a very important parameter for assessing rock durability; other tests such as those suggested by Wakeling (ref. 16) and Fookes & Poole (ref. 1) appear to be appropriate for durability assessment but need careful interpretation and should be considered as a group since results from a single test may be misleading.

39. The dolerite tested (key no. 8 in table 5) is the most obvious example of this. The rock has a K_{1c} value of 1.4 $MN/m^{-3/2}$ and an AIV of 9.9. Both of these values suggest that it is of acceptable quality. However, with a sulphate soundness loss of 36.5 weight percent, it is obviously not nearly so acceptable. Hence, it is very important that not only fracture toughness but the other engineering values are met. Only then is it possible to screen for rocks which may be weathered or highly porous. The values suggested in the literature for the various tests as criteria for satisfactory breakwater material are given in Table 6. As a result of the current study, modification to some of the published values are suggested and are also included in Table 6. Comparison of the values given by the various authors in the table suggest a progression of refinement of specified acceptable test values in parallel with increasing knowledge arising from studies into the

evaluation of the durability of rock materials in the marine environment.

Table 6. Suggested test values for armour stone.

Test	Old Recommended value 1977 (ref 16)	K_{1c}	Old Recommended value 1981 (ref 1)	K_{1c}	New recommended value	K_{1c}
Aggregate Impact Value	30 max	0.20	>20 (ACV)	0.42	16 max	0.54
Magnesium Sulphate Soundness	18% max	0.70	≯8	1.0	12 max	1.0
Water Absorption	3% max	0.54	≯2.5%	0.72	2.5% max	0.72
Apparent Relative Density	2.6 min	0.76	≮2.6	0.70	2.6 min	0.70

SUMMARY AND CONCLUSIONS

40. The size and shape of primary armour blocks is initially largely a function of the discontinuity spacing present in the rock at the quarry face. Careful blasting techniques coupled with careful handling and placing involving a minimum number of handling steps will maximise the number of blocks available for primary armour. Care in placing can also improve the interlock characteristics of the blocks and control the void ratio of the armour layer.

41. Once in place the blocks slowly become modified with time as a result of abrasive and fracture mechanisms; effects which are most important in the intertidal zone of the breakwater. These ageing effects are important in that they modify both shape and weight of the blocks and hence interlock and void ratios of the armour layer. Such modifications will gradually change the design characteristics of the structure. The study suggests that it is possible to predict the rates at which these changes take place, for given rock types in given breakwater environments, and therefore it should be possible to prepare designs which require a specified rock life by taking account of potential changes which will take place during the service life of the structure.

42. Factors which are of importance in considering the expected service life of rubble mound structures must include quality assessments of the rock materials used, the meteorological and wave climates and storm frequency. The rock quality is best assessed by the series of tests summarised in Table 6. Meteorological climatic conditions have a relatively small effect on rock durability in a marine environment except in certain special cases; for example freeze-thaw in cold climates most adversely affects rocks which contain many microfractures thus the test results on fracture toughness would be of greatest significance. In contrast in hot arid environments salt attack is more important

than in other climates so that more attention should be given to soundness test results. Table 7 gives a very preliminary summary of expected rock durability in different meteorological climates based on the principal areas of the study.

Table 7. Rock deterioration expectancy in different meteorological climates.

Determination & rock type / Met. climate	Abrasion Rounding				Spalling				Catastrophic Failure			
	A	B	C W	C S	A	B	C W	C S	A	B	C W	C S
Freezing winters	2	2	5	2	2	3	3	2	2	2	3	2
Temperate (ie UK)	3	3	4	3	1	2	2	1	1	1	3	2
Hot dry (ie Arabian Gulf)	3	4	5	4	2	3	3	2	2	2	4	3
Sub-tropical (ie E.Australia)	4	5	4	3	3	4	2	1	1	1	3	2

A = 'Acidic' rocks e.g. Granite family, Andesite family, Sandstones, Gneiss
B = 'Basic' rocks e.g. Basalt family, Andesite family, Schists, Greywackes
C = Carbonate rocks e.g. Limestones, Marbles, Dolomites
S = Strong, e.g. Carboniferous limestone
W = Weak, e.g. ME limestones, chalks
1 = Very high resistance to deterioration
2 = High resistance to deterioration
3 = Moderate resistance to deterioration
4 = Poor resistance to deterioration
5 = Very poor resistance to deterioration.

43. Wave climates and storm frequency are of course of great importance when considering structural deterioration. Gradual changes in shape and weight of blocks occur under normal sea conditions but it is storm conditions that initiate damage that can sometimes lead to rapid deterioration of the structure. The pattern of damage is produced as a result of the energy of the wave climate; the 'quality' of the rock armour, whether for example incipient flaws are present which can lead to fracture, and the 'quality' of placement on the structure with good interlock and void ratio being important.

44. Insufficiently detailed damage assessments have been made to date to draw more than the broadest conclusions and the need to study changes in damage pattern with time is important in future studies. It would however be reasonable to correlate high percentages of cavities in the damage assessment with high energy wave climates and with poor interlock or undersized material. Unstable blocks are the most prevalent type of failure type and again may be correlated with poor interlock, but in high energy wave climates tend to be removed leaving cavities.

ACKNOWLEDGEMENTS
45. The authors wish to express their grateful thanks to the Hydraulics Research Station, Wallingford and the Science & Engineering Research Council for financial support of the study. They are also grateful for much support and helpful discussions with many engineers and managers and particularly Mr. H. Price and Mr. W. Allsop of H.R.S. Wallingford, members of the Dubai office of Halcrow International Partnership and Mr. D.D. Coffey, Mr. W. Bremner and Mr. A.W. Smith who gave considerable help to members of the team while in Australia. Prof. Fookes gratefully acknowledges help and support from Rendel, Palmer & Tritton, London.

REFERENCES
1. FOOKES, P.G. & POOLE, A.B. Some preliminary considerations on the selection and durability of rock and concrete materials for breakwaters and coastal protection works. Quar.J.Eng.Geol. 14 1981, 97-128
2. DIBB, T.E., HUGHES, D.W. & POOLE, A.B. The identification of critical factors affecting rock durability in marine environments. Quar.Jl.Eng. Geol. 16 1983, 149-161.
3. DIBB, T.E., HUGHES, D.W. and POOLE, A.B. Controls of size and shape of natural armourstone. Quar.Jl.Eng.Geol. 16 1983, 31-42.
4. HUDEC, P.P. Rock weathering on the molecular level. Eng.Case Histories No. 11 1978, 47-51.
5. DARBYSHIRE, M. & DRAPER, L. Forecasting wind generated sea waves. Engineering 195, 1963, 482-4
6. FOOKES, P.G., DEARMAN, W.R. & FRANKLIN, J.A. Some engineering aspects of rock weathering with field examples from Dartmoor and elsewhere. Quar.Jl.Eng.Geol. 4 1971, 139-185.
7. BRITISH STANDARDS INSTITUTION. Site investigations. London 1981. BS 5930.
8. KRUMBEIN, W.C. Measurement and geological significance of shape and roundness of sedimentary particles. Jl.Sed.Petrol. 11 1941, 64-72.
9. SPERLING, C.H.B. & COOKE, R.U. Salt weathering in arid environments. 1. Theoretical considerations. Papers in Geography N . 8 Bedford College, London 1980, 4-46.
10. AMERICAN STANDARD TESTING MATERIALS. Soundness of aggregates by the use of sodium sulfate or magnesium sulfate C88-76, 14, 48-53. Philadelphia PA.
11. CHAPMAN, R.W. Salt weathering by sodium chloride in the Saudi Arabian desert. Am.J.Sci. 280 1980, 116-129.
12. AMERICAN STANDARD TESTING MATERIALS. Resistance to abrasion of large size coarse aggregate by use of Los Angeles machine C535-69, 14, 1975 331-332, Philadelphia PA.
13. AMERICAN STANDARD TESTING MATERIALS. Standard test method for plane-strain fracture toughness of metallic materials E399-78a 10, 1979. 540-561. Philadelphia PA.
14. BRITISH STANDARDS INSTITUTION. Methods for sampling and testing mineral aggregates, sands and fillers. London 1975 BS 812 4 parts.
15. BROCH, E. & FRANKLIN, J.A. The point load strength test. Int.Jl.Rock Mechanics Min.Sci. 9, 1972, 669-697.

16. WAKELING, H.L. The design of rubble breakwaters. Symp. on design of rubble-mound breakwaters.   Paper No 5. Experimental & Electronic
Laboratories. British Hovercraft Corp. 1977 18pp.

17. HOSKING, J.R. & TUBEY, L.W. Research on low
grade and unsound aggregate. TRRL Report LE 293
1969.

4   Concrete and other manufactured materials

A. R. CUSENS, BSc, PhD, FICE, FIStructE, FRSE, University of Leeds

The main problem of application of concrete to breakwaters is in precast concrete armour units. Plain unreinforced concrete is traditionally used and the strength of the unit is thus dependent upon the tensile strength of concrete. The paper discusses the tensile strength and durability of plain concrete and also considers the use of polymer concrete and fibre reinforcement. New technical advances in modified concrete are described as a possible key to the future. The problems of reinforced concrete in sea walls are also discussed.

INTRODUCTION

In the construction of breakwaters the predominant material is natural rock or concrete. Rock is excluded from consideration in this paper and mention of materials other than concrete will be dealt with in a few lines only. Thus despite its title, this paper is concerned almost wholly with concrete.

Referring to the typical cross-section of a rubble-mound breakwater shown in Fig. 1, concrete is principally used in the armour layer. Each armour unit is precast (normally of plain, unreinforced concrete) to a shape determined by the designer in response to a variety of hydraulic, structural and cost requirements. The unit has to protect the main breakwater and also to reduce the wave energy, both incident and reflected. The unit must be simple (and cheap) to cast and to handle; it must have a low centre of gravity, be able to withstand extreme wave conditions and be stable in the breakwater in terms of consolidation and settlement. It should be easy to repair and lead to a stable design of breakwater at crest and toe. Units are either placed according to a pattern or randomly positioned and there are probably fifty or so different designs of concrete armour unit; no doubt these will be listed and described by others at this conference.

In the construction of armour units the use of plain unreinforced concrete still predominates. Conventional steel reinforcement or pre-stressing with steel are regarded as uneconomical measures. However there has been limited use of chopped fibres of steel or poly-ethylene in armour units. Polymer-modified concrete has not been used but may be a practical proposition in the future. The paper also looks at recent developments in the production of high strength cement products which point the way to new technology in this field.

In early breakwater construction timber was employed but this is no longer the case. Bitumens are used for sealing joints and do not create any problems specific to breakwaters. Filter fabrics are included in the form of woven nylon nets or porous mats in the filters which separate different grades of rock. These are described in more detail by Fookes and Poole in their paper.

Alternatively concrete caissons have been used as breakwater and seawall units. Gerwick (ref. 1) has described caissons with walls perforated, used at Baie Cemeau in Quebec, to absorb wave energy as with the prestressed Ekofisk storage caisson. Prestressing has significant advantages over conventional reinforcement in sea walls in that it provides greater crack resistance and durability. The problems of corrosion of reinforced concrete walls are discussed in the paper.

PLAIN CONCRETE

Concrete is a material which possesses an intrinsically high compressive strength and it is this property which has led to its wide application in civil engineering projects. However it is a brittle material which exhibits a low tensile strength, far lower than might be expected on the basis of molecular cohesion. As Neville (ref. 2) explains, this is due to the occurrence of flaws and the consequent order of tensile strength is only about 0.05% of its theoretical value of 10,000 N/mm². In structural members the tensile zone of the concrete section is conventionally reinforced with steel bars or prestressed with high tensile steel strand or rods but for breakwater units it is normal practice to use unreinforced concrete. Thus the critical design parameters for these units are the tensile strength and durability of the concrete.

Tensile strength

In general the quality of hardened concrete is assessed by the crushing strength of cube or cylinder specimens at a standard age (normally 28 days). Tensile strength is less commonly assessed and test results are obtained in three ways:

Breakwaters—design and construction. Thomas Telford Ltd, London, 1984

43

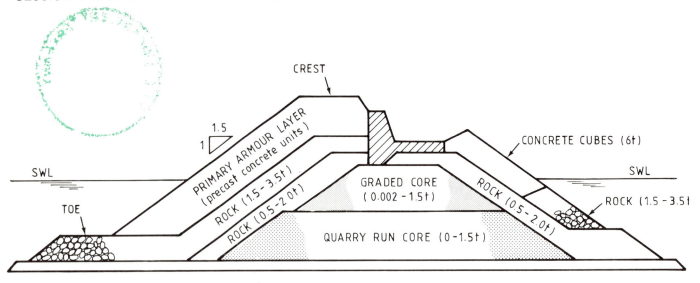

Fig. 1. Typical breakwater construction

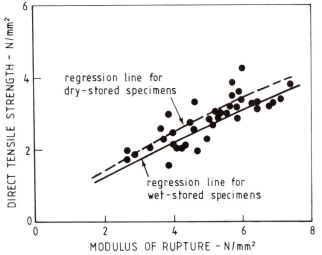

Fig. 2. Relation between direct tensile strength and modulus of rupture (Brooks & Neville, ref. 3)

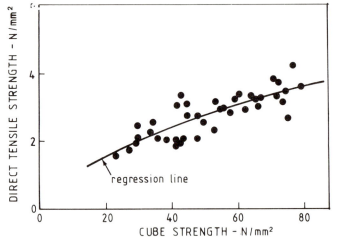

Fig. 3. Relation between direct tensile strength and cube crushing strength (Brooks & Neville, ref. 3)

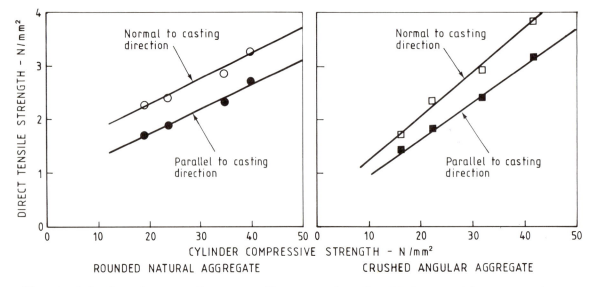

Fig. 4. Relations between direct tensile strength and cylinder crushing strength showing influence of aggregates and casting direction

a) Flexural tests of beams (usually third point loading) giving a value of *modulus of rupture*

b) Diametric splitting of a cylinder giving a value of *tensile splitting strength*

c) Direct tensile tests using special shaped specimens or special gripping devices leading to a value of *direct tensile strength*

All three methods of test give a tensile strength value but the method has a significant influence on the result. Thus modulus of rupture values tend to be somewhat higher than tensile splitting or direct tensile results. This is illustrated by Fig. 2 giving comparative results by Brooks and Neville (ref. 3).

In terms of the performance of concrete armour units it appears realistic to consider the tensile strength in bending, i.e. the modulus of rupture, as being the appropriate mode of likely failure in service.

The tensile strength is related to the cube strength of the concrete although the relationship is dependent upon the aggregate type and grading and upon the condition of the specimen at the time of test. Figure 3 shows results for a range of mixes using natural and lightweight aggregates, various water/cement ratios and admixtures. Various attempts have been made to formulate a simple empirical equation and a power law of the type $f_t = k\, f_c^n$ has been favoured. Various values of the coefficients k and n have been suggested on the basis of experiment. The author suggests the equation

$$f_t = 0.8\ \sqrt{f_{cu}}$$

as providing a conservative estimate of the modulus of rupture f_t.

The tensile strength is improved by the use of cracked, angular aggregates and the value also appears to be affected by the direction of casting the concrete. These factors are illustrated in Fig. 4 which is prepared from results recently published by Fenwick and Sue (ref. 4). The mean size of aggregate also affects the value of tensile strength (Fig. 5).

The stress-strain relationship for concrete in tension is almost linear and the occurrence of further strain beyond the point of maximum stress can be discounted for practical purposes (Fig. 6). There is little or no reserve of ductility available to help to resist impact loading.

The flexural strength in tension is decreased by repetitions of load and Neal and Kesler (ref. 7) suggest that for 10 million cycles the fatigue strength is 55% of the static value.

The onset of failure in tension is accelerated by the occurrence of large pores or flaws in the concrete. Clearly surface cracking caused during the production or handling of the armour

units must be avoided. In large volume units the heat of hydration can cause thermal strains due to differential temperature effects between the interior and the surface of the unit. These can be reduced by the use of insulated formwork or by employing a low heat cement such as slag cement or cement with pulverised fuel ash.

Careful protection of the units at early ages can prevent cracks due to plastic shrinkage or initial drying shrinkage, which are particular dangers in hot climates. It is common practice to strip the upper part of the formwork of anchor units for early re-use and unless this is accompanied by adequate covering of the green concrete, surface cracking is almost inevitable. In large units special precautions are necessary to allow for shrinkage in the mould. In some dolos moulds rubber gaskets have been incorporated but these are not always adequate and parts of the mould may have to be released within 12 hours of casting.

Durability

The ability of concrete to survive in its working environment for its design life is broadly dependent upon its quality and upon the severity of the environment.

A 'quality' concrete is usually one possessing high strength but in the context of armour units it is important that the concrete possesses a high resistance to penetration by salts in solution. In other words the concrete must have a low permeability, and to this end a low water/cement ratio, good compaction, careful curing and dense, impermeable aggregates can all play a part.

In specifying concrete the cement content and water/cement ratio are more important than the characteristic cube strength. In terms of CP110, for example, it has been shown that the specified strength grades are set too low to satisfy other specified requirements for durability. In consequence a characteristic strength of 45 n/mm² should be regarded as essential for concrete for breakwaters.

The main danger to concrete in breakwaters is attack by sea water due to the presence of sulphates. These react with the cement to form gypsum and calcium sulphoaluminate which have higher volumes than the original compounds. This can lead to expansion and eventual disintegration of the concrete. Additionally salts can crystallize in the pores of the concrete and thereby create internal pressures and further disintegration. The alternate wetting and drying endured by breakwaters enhances the severity of these attacks. In tropical climates the effects will be even more serious and in cold climates frost action will accelerate the disruption of the concrete.

POLYMER IMPREGNATED CONCRETE

There was a surge of interest and an accompanying flurry of research activity in the field of polymer concretes in the decade 1965-1978. Few practical applications have followed

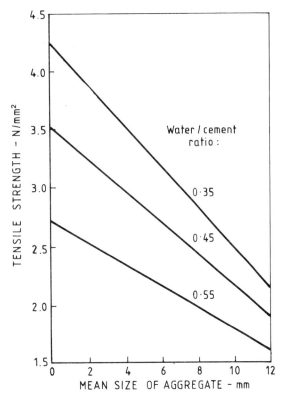

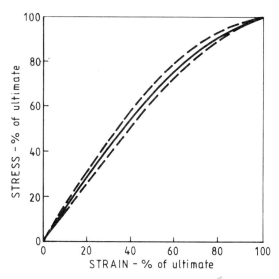

Fig. 5. Relation between direct tensile strength and mean aggregate size (Johnston, ref. 5)

Fig. 6. Stress-strain curve for concrete in tension (Domone, ref. 6)

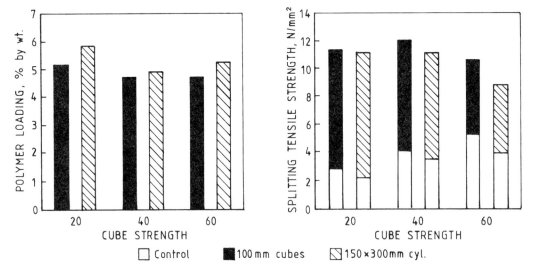

Fig. 7. Effect of concrete quality upon polymer loading and tensile strength (Sopler *et al.*, ref. 8)

presumably for economic reasons. Even the partial impregnation of concrete specimens can create a substantial increase in the tensile strength properties of plain concrete specimens. Sopler, Fiorate and Lenschow (ref. 8) report that soaking of dry concrete specimens in methyl methacrylate monomer (MMA) in a cold storage chamber at temperature in the range 0°C-5°C followed by polymerization in a bath of warm water. (70°C) for four hours led to substantial improvements in tensile (cylinder splitting) strength (Fig 7) and also in cube compressive strength.

FIBRE REINFORCED CONCRETE
The use of steel fibres in concrete creates a marginal increase in the modulus of rupture to initial cracking but it is suggested by Hannant and Edgington (ref. 9) that in a marine environment there is little likelihood of the corroding fibres enhancing tensile strength for an extended period of time. However a recent paper by Godfrey (ref. 10) reports that 40 tonne dolosse cast in 1972 using concrete reinforced with steel fibres (cylinder strength 59 N/mm²) for breakwaters at Humboldt Bay, Eureka in California had experienced few problems; only two or three had broken. On the other hand, many of the unreinforced dolosse used in the same construction had fractured in the ten-year period.

Polypropylene fibres have also been used in concrete, principally in piling units to improve impact resistance. Polypropylene fibres lack the strength of steel fibres but eliminate the possibility of corrosion and they were employed in the SHED concrete armour units recently installed in the new seawall at St. Helier in Jersey. The fibres were reported (ref. 11) to lee 50 mm in length and the total fibre content of the concrete mix was 0.2% by weight. The basic quality of the concrete (28 day cube strength reported as 32 N/mm²) was not high but the tensile strength would have been enhanced slightly, impact strength improved and brittleness reduced.

The inclusion of fibres can create problems in that the fibres tend to ball up in the mixer. This can be solved by careful addition of the fibres to the mix. The workability of the mix can also be affected adversely and vibration is essential for concrete placing.

MODIFIED CEMENTS
In 1969, Lawrence (ref. 12) reported experiments in which he had compacted dry cement at high pressures and subsequently hydrated the compacted material. The resulting hardened cement pacts exhibited a high compressive strength up to 374 N/mm² and a tensile strength of 15 N/mm²; failure in tension was described as very brittle.

Bache combined the benefits of superplasticized concrete (i.e. concrete containing an effective surfactant agent to disperse Portland cement in water) with the addition of very fine particles (mean diameter 0.1 μ) of silica fume to fill the spaces between the cement particles.

Mortars and concretes made with this material (known as DSP - Densified Systems containing homogeneously arranged ultra-fine Particles) are reported (ref. 13) to give high cylinder crushing strengths (125 to 270 N/mm²) and possess excellent resistance to freezing and thawing. However they are said to be very brittle and it is recommended that some form of reinforcement be used to provide ductility.

Birchall, Howard and Kendall (ref. 14) have developed macro-defect free (MDF) cements by dosing neat cement paste of low water-cement ratio with surfactant agents. After mixing, the stiff paste is compacted to produce an air-free substance for extruding or forming into shape. These new materials can reach tensile strengths in flexure of 70 N/mm² although the failure is brittle in nature. The authors now claim to be producing a second generation of MDF cements by modification of the microstructure after removal of defects where the flexural tensile strengths of 150 N/mm² are being achieved.

All of this work is currently only applicable to very small components. However, given time these technological advances will be developed for larger scale elements and should eventually lead to practical unreinforced concrete with a considerably enhanced tensile strength - precisely what is needed for armour units.

REINFORCED CONCRETE
The severe environmental factors experienced by reinforced concrete in sea walls and breakwaters make it essential that the highest standards of design and construction are followed. If standards are less than high, the design life of the structure is very short - in extreme cases extensive repairs are necessary after only a year or two of service. The durability of plain concrete has already been referred to; the general durability problem is exacerbated by the pressure of steel reinforcement. Gerwick (ref. 1) has pointed out that a piece of exposed steel situated in the inter-tidal zone will corrode less than the same steel embedded in highly permeable concrete in the same environment. If sea water permeates through the concrete, chloride deposits are set up in the pores and these establish galvanic cells with the steel reinforcement leading to electrolytic corrosion. Alternate wetting and drying accelerates the process and expansion due to the corrosion products spalls off the concrete cover.

The factors affecting general durability: impermeability, low water/cement ratio, high cement content, good compaction and controlled curing all remain of importance. In addition adequate concrete cover is essential and 60-75 mm is the normal specification range for reinforced concrete. However lower values have been used successfully with high quality concrete; interestingly in ferrocement boats a cover of 10 mm appears to be adequate.

The adoption of prestressed concrete in seawall and breakwater construction is recommended wherever feasible. The sustained compressive stress on the structure prevents the formation of cracks in the concrete and reduces the likelihood of corrosion.

The feasibility of use of non-corrosive reinforcement consisting of die-drawn polyethylene rods with tensile strengths comparable to steel is currently being investigated at the University of Leeds with a view to applications in coastal and marine structures.

DISCUSSION AND CONCLUSIONS

The practice of using plain concrete for breakwater armour units is understandable in terms of economics but in the author's view is difficult to justify in engineering terms in most circumstances. In particular the loading on randomly placed units due to waves and impact from neighbouring units will easily create tensile stresses above the intrinsically low ultimate values exhibited by plain concrete. The introduction of fibre reinforcement will clearly help to some extent and it is surprising that its adoption·has not been more widespread. With pattern-placed units, the loading is more predictable but particular designs of unit may be suitable only for small breakwaters.

It appears that engineers who have to design and construct breakwater units are having to work ahead of technological developments. There is a clear need for more information on the loading of units in service and for a concrete possessing a significantly higher tensile strength than plain concrete. However the quality of concrete conventionally employed for precast armour units does not appear to be exceptionally high and there is little evidence of great care being employed in the casting and curing of these units.

The author's general conclusions are:

1. The use of plain reinforced concrete armour units in breakwaters poses severe problems. Some reduction of the current level of risk may be achieved by:

   a) the use of higher quality concrete and improved casting and curing procedures

   b) the inclusion of fibre reinforcment in the concrete

2. New technological advances in modified cements and polymer concretes seem unlikely to affect engineering practice in armour units in the near future.

3. There is an urgent need for research data on the stresses in service of concrete armour units.

4. In applications of reinforced concrete in sea walls more attention also should be paid to concrete quality and to standards of placing and curing. Wherever prestressed concrete is a feasible proposition it is to be preferred to normal reinforced concrete.

REFERENCES

1. GERWICK B.C. Marine Structures. Handbook of Concrete Engineering (Ed. M. Fintel), 1974, Van Nostrand Reinhold, 615-630.
2. NEVILLE A.M. Properties of Concrete. Pitman, London, 1973.
3. BROOKS J.J. and NEVILLE A.M. A comparison of creep, elasticity and strength of concrete in tension and in compression. Magazine of Concrete Research, 1977, 29, 100, September, 131-141.
4. FENWICK R.C. and SUE C.F.C. The influence of water gain upon the tensile strength of concrete. Magazine of Concrete Research, 1982, 34, 120, September, 139-145.
5. JOHNSTON C.D. Strength and deformation of concrete in uniaxial tension and compression. Magazine of Concrete Research, 1970, 22, 70, March, 5-16.
6. DOMONE P.L. Uniaxial tensile creep and failure of concrete. Magazine of Concrete Research, 1974, 26, 88, September, 144-152.
7. NEAL J.A. and KESLER C.E. The fatigue of plain concrete. Proceedings of the Conference on the Structure of Concrete, 1968, Cement and Concrete Association, 226-237.
8. SOPLER B., FIORATO A.E. and LENSCHOW R. A study of partially impregnated polymerized concrete specimens. 1973, Polymer in Concrete, Publication SP40, American Concrete Institute, 149-171.
9. HANNANT D.J. and EDGINGTON J. Durability of steel fibre concrete. 1975, Rilem Symposium on Fibre Reinforced Cement and Concrete, Construction Press, 159-170.
10. GODFREY K.A. Fiber reinforced concrete. Civil Engineering - ASCE, 1982, 52, 11, November, 44-50.
11. ANON. New concrete armour unit tried on seawall on the island of Jersey. ASCE News, 1982, April, 11.
12. LAWRENCE C.D. The properties of cement paste compacted under high pressure. Research Report 19, 1969, Cement and Concrete Association.
13. BACHE H.H. Densified cement/ultra-fine particle based materials. Paper presented at 2nd International Conference on Superplasticizer in Concrete, 1981, June (to be published by American Concrete Institute).
14. BIRCHALL J.D., HOWARD A.J. and KENDALL K. New cements - inorganic plastics of the future. Chemistry in Britain, 1982, December, 860-863.

Discussion on Papers 3 and 4

A. B. POOLE, P. G. FOOKES, T. E. DIBB and D. W. HUGHES (*Introduction to Paper 3*)
It is necessary to underline various comments in the Paper concerning rounding of armour stone (and probably of stone beneath the outer layer, although we have no data as yet concerning these). The principal mechanism producing rounding in the life of the armour when it is interlocked and not free to be rolled around by wave action, is by abrasion (i.e. loss of stone angularity by wave attrition). Spalling and catastrophic fracture also help reduce rock weight and reduce interlock between individual blocks. The end product of this action is that individual stones may eventually become free to be rolled by heavy seas. This in turn may lead to the breakup of the breakwater and possible development of a low stable angle slope of rock rubble which may then act like an energy spending beach.

Fig 1 illustrates the relationship between roundness of interlocked blocks measured on actual breakwaters (cf. Fig. 10 in the Paper) and other types of damage expressed as a damage coefficient. The damage coefficient is defined here as the proportion of unstable or substandard blocks (rounded, degraded, fractured), expressed as a percentage of the total number of blocks in a given study area (usually 100 m^2).

Reductions in armour stone interlock due to rounding will affect the structural stability of the breakwater but at present there is no satisfactory means of defining or assessing the interlock factor. Such considerations are normally treated as subjective assessments based on experience, and a means of quantifying interlock is therefore a main requirement of our research.

All current interlock factors are derived from laboratory model studies which are intended to study rounding effects by wave abrasion. Typically damage is assessed solely by the movement of armour units (rolling by wave action) or the instability of individual blocks in the model. These methods are useful as basic guides to allow the derivation of more satisfactory techniques in which different types of deterioration process may be weighted in terms of their relative importance. Such studies are currently being undertaken by Queen Mary College on full-scale breakwaters and as scale model experiments.

A monitoring programme was begun early in 1982 and with the limited data available at present, it is only possible to compare the progression of wave attrition rounding with the crude assessment of damage coefficients and

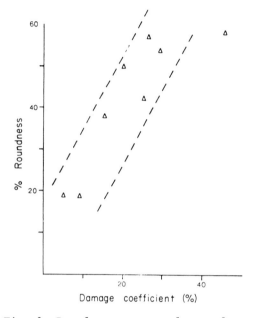

Fig. 1. Roundness expressed as a damage coefficient

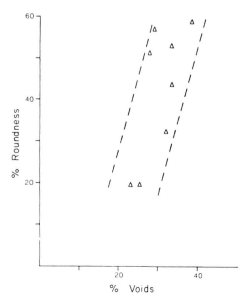

Fig. 2. Roundness plotted against voids

Breakwaters—design and construction. Thomas Telford Ltd, London, 1984

49

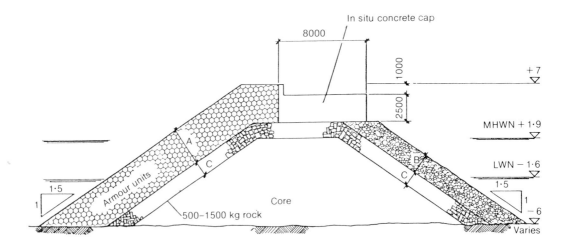

A Armour units 5·5t dolosse or equivalent
B 1500–2500 kg rock
C 500–1500 kg rock

Fig. 3. Typical breakwater cross-section

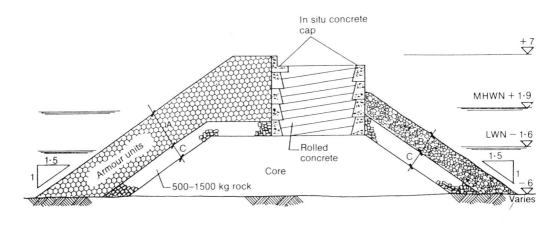

A Armour units 55 tonne dolosse or equivalent
B 1500–2500 kg rock
C 500–1500 kg rock

Fig. 4. Typical breakwater cross-section, using rolled concrete

void ratios of armour layers. Both of these are related to armour block stability and the data currently available is plotted on Fig. 2. The void ratio is defined here as the percentage void area to percentage armour area over 100 m^2 of breakwater face.

The design engineer will no doubt wish to predict the average reduction in weight of the armour due to wave attrition processes over the design life of the structure. It should be possible to make predictions concerning abrasion weight loss with time, on the basis of experimental or standard test data. The Los Angeles Abrasion value, for example, might be compared with known performance of rock types,which would allow ranking of armour stone on the basis of relative wave attrition in the particular

marine zone and local environment. Fig.10 in the Paper, based on field data collected from similarly designed breakwaters in similar environments and constructed with similar carbonate rocks, illustrates real rounding with real time. Figs 11 and 12 in the Paper are simulated experiments in the laboratory.

R. W. J. BOWRING (PSA, Plymouth)
The eastern mole at Gibraltar is a composite breakwater built of limestone. The lower courses of armouring blocks on the seaward. side have in many cases been colonized by a boring mollusc of the mussel family. As the mollusc grows (to about 100 mm long by 70 mm dia.) the strength of blocks are gradually eroded leading to failure of the breakwater. This is overcome by replenishment. Are there any other known instances of breakwater damage by molluscs?

E. H. HARLOW (Consultant, Texas)
Little has been said so far about core materials which represent about 80% of the material in a large rubble-mound breakwater, especially when one considers that the specific surfaces of these smaller pieces of rock increase exponentially in inverse ratio to their sizes. These surfaces offer vastly greater opportunity for weathering and degradation, making each particle still smaller. Is there any data on what influence this process has upon the configuration and outline of the interior mass on which the underlayers and armour rest?

B. L. FLOWER (Associated British Ports, Swansea)
My responsibility includes the maintenance of
four breakwaters. During the construction of
the main breakwater at Port Talbot (1966-70) the
6-8 ton armour stone had come from at least 30
quarries, some from as far away as Derbyshire.
Unfortunately, some of the limestone armour
showed deterioration after about five years and
in 1976 repairs were carried out using 20 ton
ten-sided concrete blocks cast in situ. Although
carefully placed, most of these had rolled down
the slope, but fortunately they now formed a
very good toe for the 11 ton tripod blocks
which had been placed since then; 500 being
placed in 1978, 850 in 1981 and a further 850
would be placed in 1983.

I have tried taking aerial photographs to
determine the amount of damage, but this has not
proved satisfactory. The most effective method
of determining the number of tripods required
to repair the damage is to average visual
observations. I think that so far approximately
$1\frac{1}{2}$% of the tripods were broken and a similar
number turned over.

The deterioration of the limestone has led me
to believe that generally it is not appropriate
to use sedimentary rocks as armour where the
weight requirement is more than about 5 tons,
and for heavier armour then either igneous rocks
or concrete units should be used.

The tripods have proved very cost effective.

K. A. KAYE (Shephard Hill Ltd)
Professor Cusens mentions the use of polypropylene fibres in the SHED armour blocks in
the new breakwater in Jersey and suggests that
the tensile strength of the concrete would have
been enhanced as a result. We did not find this
to be so. From some limited comparative tests
which we carried out on site, we found that in
the case of compression tests based on 28 day-
cube strengths there was a reduction of 10% in
the case of fibre-reinforced concrete compared
with unreinforced concrete. In the case of
comparative tensile strengths based on cylinder
tests we found no discernible difference.

The main advantage of using fibres seems to
be in the increased resistance to impact loading
stresses achieved. Some useful work on this
was carried out at the University of Surrey by
Hibbert and Hannant (ref. 1).

The additional cost of using polypropylene
fibre reinforcement in a 2 t SHED block is only
about £6 which seems a worthwhile investment.

We experienced no problems with balling
while using polypropylene fibres although we
had difficulties when using steel fibres. We
also found it impossible to prevent some of the
ends of the steel fibres protruding from the
concrete on unformed faces and at some joints
in the formwork. One could therefore not contemplate using steel fibres in any location
where children can climb over the units.

Another thing which we do almost as a matter
of course on all our contracts is to replace
some of the cement in the mix by PFA. This was
also done in the case of the SHED blocks in
Jersey. We find this increases both the density
of the concrete and the workability; (alternatively the water/cement ratio can be reduced for
the same workability) and it produces a better
finish. Resistance to sulphate attack is also
increased and the danger of aggregate/alkali
reaction is reduced. I can never understand
why one does not hear more of the use of PFA,
especially for marine structures, at least in
and around the UK when there is a reliable
supply of consistently good quality PFA always
available.

P. LACEY (Ove Arup & Partners)
I would like to make a contribution concerning
new materials for breakwater construction.
Fig. 3 shows a typical breakwater cross-section
for Eastbourne Harbour designed in 1975. In
1979 the scheme was reappraised and at that
time rolled concrete was being used in dam
construction.

Fig. 4 shows what the cross-section could
look like if rolled concrete is used. The
advantages are speed of construction and very
dense concrete which can be overtopped minutes
after laying without harm. The outside kerb
is concrete placed by slip form. The savings
due to the small plant requirement and speed
of laying amounted to some 25% of the estimated
total cost of the breakwater.

What is needed now is a prototype test
section in marine conditions.

J. F. MAQUET (Sofremer)
The quality of concrete should be adapted to
the type of units involved. A strength of
45 N/mm^2 can be necessary for dolosse which
suffer tensions and flexions. A strength of
30-35 N/mm^2 is frequently accepted for tetrapods
depending on the size. As far as cubes are
concerned, the main loads are of a compressive
nature and Antifer cubes have achieved a compressive strength of 25 N/mm^2. Examples of
units with this strength show a good durability
because the effects of loads are not as detrimental as for other units.

H. F. BURCHARTH (Aalborg University, Denmark)
Results from full-scale static and dynamic
(impact) tests with dolosse in the range up to
30 t can be summarized as follows:

A low content of steel fibre reinforcement
(50-70 kg/m^3) will not improve the static and
dynamic strength compared with the strength of
unreinforced units (i.e. the onset of damaging
cracks starts at the same load/energy level).
Conventional steel bar reinforcement is superior
to steel fibres of the same amount. By applying
100-130 kg steel bars/m^3 concrete, spalling and
not cracking seems to be the limiting factor in
the case of impact loading, even for units of
30 t.

In the literature it is often regarded that
the impact strength is increased a lot even if
a small amount of fibres are added. This statement is based on a fall-hammer test where the
energy to split (completely disintegrate) the
specimen (i.e. a beam) in one blow is calculated. Such tests (including the results) are
not relevant to armour units because the failure
criterion for armour units is the development
(the onset) of cracks with a small width.

Fibres are useful mainly in very slender
flexible members, not in stiff elements like
dolosse, tetrapods, etc.

Full-scale impact tests with high strength (densified) concrete dolosse showed that nothing is gained in impact strength (because such concrete structures are very brittle). The long-term durability is of course improved.

O. T. MAGOON (*US Army Corps of Engineers*)
The projection of steel fibres in fibre-reinforced concrete dolos armour units are not causing spalling of concrete. They provide a rather rough surface and may cause concern if used for units where there is large public use.

Steel reinforcement has been used satisfactorily in the Humboldt dolos units. Reinforcement may be practicable in many coastal armour units.

A. B. POOLE, P. G. FOOKES, T. E. DIBB and
D. W. HUGHES (*Paper 3*)
In reply to Mr Bowring, no significant cases of deterioration of breakwater stone were seen during the studies for our research. However, a number of organisms are known which will bore or erode rocks; these include various bivalves and echinoderms. Chalk can commonly be seen to be bored around the southern and eastern coasts of the UK as it is both relatively soft (i.e. weak) and a limestone, a favourite rock type for calcareous shelled organisms. Elsewhere we have seen boring molluscs attacking limestone, used as armour in the Gulf, and also concrete (i.e. a man-made lime rich rock) piles, particularly in the intertidal and immediately subtidal zones, but not to an extent where remedial treatment is yet necessary.

With regard to Mr Harlow's query our current research has been specially concerned only with the outer rock armour where the processes of degradation are liable to be the most severe. No field work has been carried out on core material so far, but in the follow-up research programme planned, it is envisaged that in our model studies simulating breakwater deterioration with time, it should be possible to provide some information concerning core materials.

A. R. CUSENS (*Paper 4*)
Mr Flower's remarks on the Port Talbot breakwater were of considerable interest and my only comment on the tripod units is that although only 1½% had fractured, even this low figure would have been improved upon if the angle between the legs had been less sharp. In general the design of armour units should avoid sharp corners which can act as stress-raisers.

Mr Kaye mentions that the use of polypropylene fibres for the SHED armour blocks did not lead to any enhancement of the tensile strength as measured by split cylinder tests. The modulus of rupture test might have been a more appropriate test and he might then have found a small increase in tensile strength but admittedly no more than perhaps 10%. He also makes the point that the impact strength is improved significantly by the presence of fibres. This point is denied by Professor Burcharth, but he accepts that durability is improved. My own opinion is that fibres will not stop cracking of armour units in flexural tension but that they will aid in arresting the propagation of cracks.

Professor Burcharth advocates the use of conventional steel bar reinforcement (and to some extent Mr Magoon is in support) but he reminds us of the problem of spalling due to corrosion. At the University of Leeds we are currently looking into the possibility of using modified polyethylene bars as reinforcement in semi-structural elements such as armour units. If this could become a practical possibility it would avoid the corrosion problems experienced with steel reinforcement in marine and coastal structures.

Mr Maquet's figures for the strength of concrete adequate for breakwater units strike me as being somewhat low. However, durability is related to cement content and impermeability rather than strength of cubes. I think that discussions of minimum strength may be somewhat counter-productive. I can only re-emphasize my view that the use of higher quality concrete and improved casting and curing procedures is essential in armour units and in sea wall construction.

REFERENCES
1. HIBBERT, A. P. and HANNANT, D. J. Impact resistance of fibre concrete. Report No.SR654. Road Research Laboratory, 1981.

Theme paper: The design process

J. E. CLIFFORD, BSc, FICE, Sir William Halcrow & Partners

INTRODUCTION

The aim of this Conference is to improve the design and construction of breakwaters and the organisers have been influenced by recent unacceptable damage to some structures. It might be conjectured that many breakwaters which have never recorded damage have been over-designed and resources have been wasted.

The design engineer's task is to find the structure which is in the right position between these two extremes, but design methods have provided him with a rather inaccurate weapon to aim at the target.

Design procedures are not static and evolution is always taking place by theory, experiment and experience. This has been particularly evident in the maritime environment in recent years.

How, therefore, does a commentary on the present design process differ from a review of the State of the Art?

I think the answer rests largely on the point of view of the commentator, remembering that the state of the art will be ahead of the application, other than in exceptional cases.

Perhaps the military analogy used earlier can be extended to the soldier in battle who uses the resources he has, when there are more modern weapons which have not yet reached him.

Ideally all good soldiers should have the latest weapons, and all good design engineers should be working at the frontiers of knowledge in their profession.

In maritime engineering we all know that very detailed descriptions of sea states are possible, and methods are available to utilise this information in engineering works. However, can the designer working on a project actually acquire all this detailed information? Are there practical limitations in applying the latest state of the art to the design process?

I think we can accept that different engineers will not have the same resources, and design procedures will differ between projects.

In giving this personal view of the design process, I have tried to cover the matters to which the design engineer will give attention throughout the design of a breakwater. The subject is extensive - and comment, which is necessarily brief, has been directed more towards practical issues than towards defining recommended procedures.

I propose to take for granted that there is much of value in the literature to act as the essential references and to limit my remarks to the general philosophy and sequence of design, dwelling for a while only on aspects which seem to be of some importance. In selecting the topics for reference I have also been guided by the contents of the papers at this Conference, generally paying less attention to what is already well covered.

The United States Shore Protection Manual (1977) (Reference 1) is a most comprehensive document and the reports of 1973, 1976 and 1980 of the PIANC Waves Commissions (References 2, 3 and 4) are other important sources of advice, opinion and detailed references.

Since these documents were published there have been many papers dealing with breakwater research, and some on project implementation, but documents describing experience of structures in service are less frequent.

Nevertheless there are, all the time, increasing references which will be available to the designer even if only a few may be found of direct relevance in a particular case; so that armed with all these documents and his own library, but above all with his experience, the designer may then set about the design of a breakwater.

* * * * * *

It is perhaps worth reflecting on boundary conditions in introducing this review of the present design process. There are boundaries of physical action on the structure, boundaries of knowledge of how the structure will behave, and a third important boundary - that of experience.

In terms of service conditions of wave action on a breakwater the designer needs to be conscious of whether he is extrapolating previous experience.

Breakwaters—design and construction. Thomas Telford Ltd, London, 1984

53

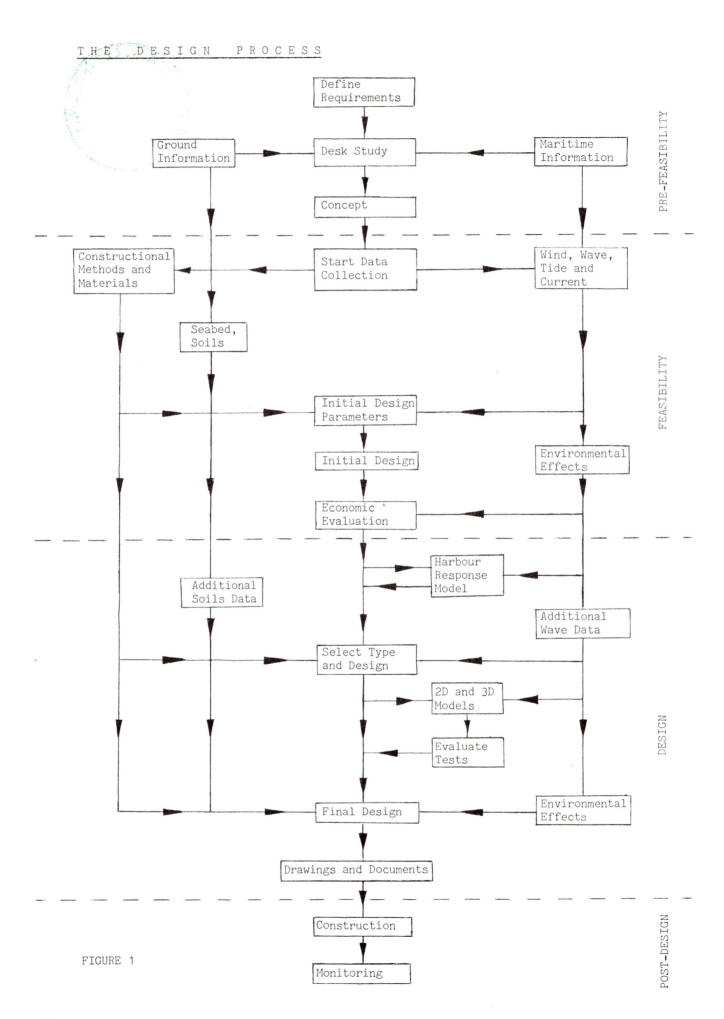

THE DESIGN PROCESS

FIGURE 1

No structure can be expected to be entirely maintenance free for, say, 50 years service, but it is possible in many cases to reduce the costs of dealing with later deterioration by spending more at the outset.

However, the whole philosophy of the design of rubble mound breakwaters in particular has been based on a certain amount of flexibility, movement and deterioration. The maintenance needed in this case is not just a function of the inadequacy of the materials or the uncertainty of their behaviour, but has been built into the basic design concept.

Whether this should be so is questionable and I suspect that most designers are now trying to limit this flexibility to a considerable degree as indicated by the heavy concrete cappings and wave walls often used.

Thus the capital/maintenance optimisations used for simple rock mounds need to be viewed with caution for large breakwaters.

Severe storms are the cause of damage, and the need for maintenance, but the timing of the exceedence of a damage threshold is unpredictable. Where a boundary condition can be defined, such as a limitation on waves due to water depth, more confidence is gained in stating an upper limit of wave action and consequent damage.

Where boundaries are less evident, for instance with deep water and long fetches, there is less likelihood that a particular severe condition will represent an extreme.

Thus boundaries are often ill-defined and the judgements the designer applies during the design process will need to take account of these limitations in any particular case.

THE DESIGN PROCESS

The design process comprises an orderly sequence of activities and decisions, and can be summarised through the progress of an imaginary project, covering layout, service conditions and design of the structure.

The design process is conveniently considered in 4 stages, illustrated in Figure 1.

The Pre-Feasibility Stage defines what is required and, with limited data, formulates a concept.

The Feasibility Stage includes a substantial data collection, initial design, and cost/benefit calculations which will determine that the scheme is practical and financially satisfactory.

The Detailed Design Stage requires further data collection. Design parameters are refined, a final design is prepared, model tests may be done, and the drawings and specifications are completed.

The Construction Stage may include changes required to the design, and concludes with specifications for monitoring and maintenance.

THE PRE-FEASIBILITY STAGE

The perceived demand for the new structure broadly describes what is required; data collection will start and plans will be made for more.

The pre-feasibility stage will usually conclude after the initial desk study of available site data and the preparation of a project concept. Commercial and economic considerations will prevail up to this time, on the very reasonable grounds that the design engineer can be relied upon to provide the technical solution.

In many cases it will be after this stage rather than earlier, that the design engineer becomes involved, but in any event it will be data acquisition which will dominate his planning at the start.

DATA COLLECTION

It is worth considering the whole question of data collection before reviewing how it is used at different stages. Data collection is required on ground and maritime information. The investigation of rock sources is often linked with the general ground investigations.

A thorough desk study would first be undertaken to assemble all the information available on the geology of the site, sea bed form and the marine climate. On the basis of the information assembled, further phases can be planned for geophysical and hydrographic investigations, boreholes and soil tests, and oceanographic data collection.

Throughout the stages of a project the data bank will continually increase and confidence in design parameters will grow.

Whatever the ground may be like, in engineering terms it is permanent, so there is no technical advantage in extending the time-scale of the investigation.

However, at the start, the precise location of the breakwater may be uncertain, so economy dictates that a sparse, but wide, coverage is appropriate. Economy is important because, although marine site investigations require conventional techniques, the costs of mounting the operation in an exposed location can be very high.

For each project it will be necessary to decide whether, to save later mobilisation costs, a single geotechnical survey should be attempted at the feasibility stage and sufficient information obtained for detailed design.

Reliability of wave statistics is important for design and for environmental effects. Storm conditions are the principal concern, but it will also be important to collect information on the "typical" conditions which must be taken into account by the designer, and later on by the contractor, in planning construction.

Given sufficient time and resources, basic wave parameters can be collected with relative ease. Accelerometer buoys offshore can achieve a high rate of useful recording, but consideration should be given to duplicating the facility, to avoid loss of data. If this is not possible, a cheaper inshore pressure recorder could be used as partial back-up, and also to give some correlation between offshore and inshore wave climates.

Wave records should be obtained for as long as possible - preferably for at least 3 years in the greater latitudes, to provide information on the frequency and duration of storms and the distribution of wave heights and periods.

Short-term wave records, however, may not be typical of the average conditions at the site, therefore longer-term visual wave data and wind data are also required for correlation against the wave records.

Often the reports of Voluntary Observing Ships are insufficient for a rigorous analysis, but an indication of the local deep water wave climate can be obtained.

Wind data enables storm frequencies to be checked and gives greater confidence in the prediction of extreme events, by the use of hindcasting techniques applied to the offshore wave generating area.

The recording of wave direction has not yet been resolved to the same degree as wave height and period measurement, although the aim must be to collect simultaneous measurements of direction and surface profile.

Where an elevated observation point is available, time-lapse radar photography can be used to record wave crest direction. Other devices record sub-surface orbital velocities but need correction for currents.

Until instruments to record wave directions at a point become well developed, offshore wave directions can be assessed in two ways. Firstly, backtracking refraction programmes from inshore visual observations; secondly by assuming that the dominant wind direction over the fetch length represents the general wave direction. The latter method is reasonably reliable over limited fetches, but where the coast is exposed to a very large fetch, use must be made of synoptic weather charts for hindcasting.

Knowledge of currents is required for breakwater alignment and arrangements for navigation of vessels. Strong current streams can also cause refraction of wave trains and may modify waves at the structure.

The collection of information on the rate and direction of littoral drift may also be important, and some predictions on the effects of the breakwater on the coastal zone will usually be required.

Tidal records may need to be collected and investigation is also needed into the occurrence of surges.

THE FEASIBILITY STAGE

At the feasibility stage the designer will be required to demonstrate both technical and economic feasibility before he is able to acquire all the data he really needs, before he can refine the designs and perhaps most important, before he can commission any hydraulic model tests.

This requires an assessment of service conditions, survival conditions and the preliminary choice of a structural solution. Available data is therefore analysed on the site conditions, construction options and the design wave climate. The wave climate includes operational conditions and particularly the design wave for structural stability.

It will be said that the concept of a design wave needs many qualifications. In terms of wave height alone, some would say it is misleading or even meaningless.

Nevertheless, it remains the single parameter which runs through past experience and gives a measure for comparison, enabling experience to be projected to a new project in its early stages.

In the present design process, therefore, the design wave is needed because it enables a start to be made. Having said that, the limitations of its use have to be recognised.

I will refer later to the structural solutions which might be considered, but the value to be assigned to the design wave height is different as between rigid structures and rubble mound breakwaters.

In the former case the designer would normally assess the maximum wave (H_{max}) likely to reach the structure over a long period.

However, in the case of rubble mound structures, the value of the design wave cannot be so confidently asserted.

Hudson's formula is still the most widely used of the expressions relating design wave height to the stability of the outer armour layer of a rubble mound. In this formula, to assess the weight of a unit of armour, it is conventional practice to regard the design wave as the largest significant wave height (H_S) expected to occur over a long period. However, an empirical stability coefficient K_D has to be chosen, appropriate to the type of armour unit.

VERTICAL-WALLED BREAKWATERS - IMPORTANT FAILURE MODES

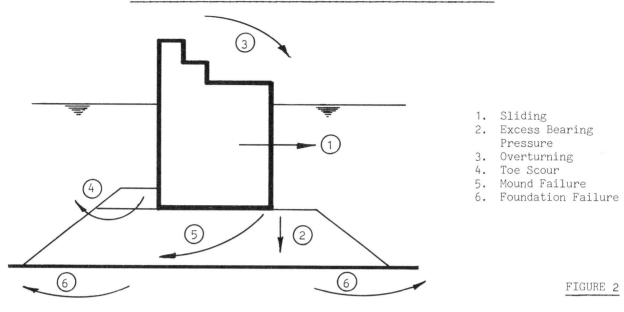

1. Sliding
2. Excess Bearing Pressure
3. Overturning
4. Toe Scour
5. Mound Failure
6. Foundation Failure

FIGURE 2

RUBBLE-MOUND BREAKWATERS - IMPORTANT FAILURE MODES

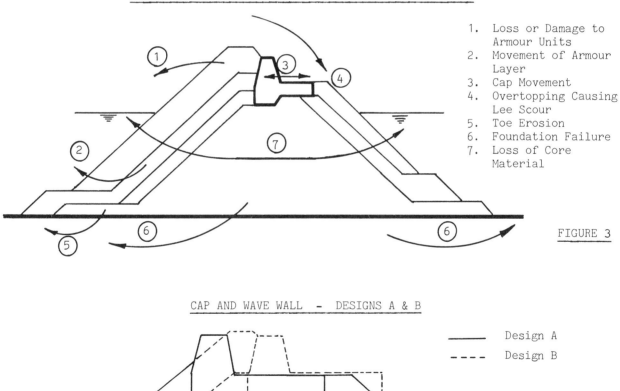

1. Loss or Damage to Armour Units
2. Movement of Armour Layer
3. Cap Movement
4. Overtopping Causing Lee Scour
5. Toe Erosion
6. Foundation Failure
7. Loss of Core Material

FIGURE 3

CAP AND WAVE WALL - DESIGNS A & B

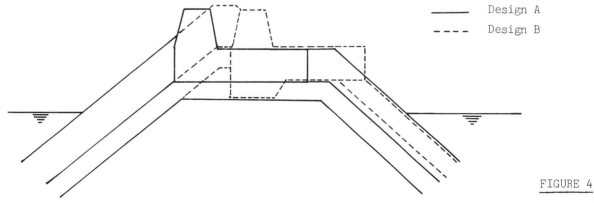

——— Design A
- - - - Design B

FIGURE 4

Many K_D values, which have been based on earlier model experiments with regular waves, are now open to considerable doubt. When considering recent experiences in prototypes and more realistic models, the inevitable conclusion is drawn that a further coefficient of caution is needed. The uncertainties arise in two ways; firstly over the correlation of prototype damage with the model tests, and secondly over the correlation of realistic wave loading with regular model waves.

This factor of caution can be introduced either by taking a lower value of K_D or by selecting a higher value of the design wave. There have been, suggestions of using H_{max}, but where H_{max} can be 1.6 to 2.0 times the significant wave height, this effectively increases the armour unit weight by a factor of 4 to 8. Other dimensions of the structure, such as the thickness of armour and underlayers, crest height and width, will also consequently increase with an appreciable effect on the cost. Most designers will conclude that to use the single highest wave for the design of a rubble mound structure is likely to be too extravagant.

Alternatively a low value of K_D can be used - perhaps half the advertised value, particularly for complex armour units, which doubles the unit weight. The same effect is achieved by multiplying the design wave height by about 1.3; that is instead of using the significant wave height, a value of about $H^1/_{10}$ could be used with the published value of K_D.

Thus, in the early design stages the designer has to make this difficult decision with serious implications on sizing the structure, whatever further studies and analyses he may do later.

Initial Design Parameters

As data collection proceeds, data analysis will become more refined with time, and the broad assumptions made at the start will become initial design parameters which must be determined when designs are commenced.

If adequate soil samples have been recovered, then laboratory tests can be undertaken to assign values to density, grading, shear strength and compressibility of the soils at proposed breakwater locations.

From data on the marine climate, an initial description of wave action for construction and service can be made. There will be insufficient information to state the wave climate in the detail which has been referred to in other Conference papers, but design waves can be assessed. These will be based on forecasting and hindcasting compared with initial plots of recorded data according to one of the recognised probability methods which gives the best fit to the records.

Many designers will consider it appropriate to adopt a 100-year return period wave condition with the height parameters of H_S and H_{max}, and with an assessment of associated wave periods. Refraction and shoaling computations will be used to transfer these waves inshore to the site to establish design wave parameters.

Initial Design

During the feasibility stage it is important to have reliable estimates of cost. Structural solutions will be investigated and probably one outline design will be selected, but the choice is not irreversible and with deeper study later on, the selection may be confirmed or varied.

From the initial design parameters, the options will be set out and each considered for its effectiveness, ease of construction and economy.

There are a number of breakwater structures to be considered but the options will generally fall into two categories; rubble mound or vertical-walled breakwaters. Although the papers at this Conference deal almost exclusively with rubble mound structures, the alternative of a vertical-walled structure must be considered and some reference to this type is appropriate.

A vertical-walled breakwater may take a number of structural forms, based generally on blockwork, infilled vertical sheet piles or prefabricated caissons.

Whatever its precise form, each type of vertical breakwater forms an impermeable barrier which reflects wave energy - although if economically designed, some overtopping is likely to occur under extreme conditions.

Few blockwork breakwaters have been built in recent years as construction is expensive and slow, and breakwaters using infilling to sheet piled walls are generally limited to modest depths of water.

Thus in the majority of applications where a vertical-walled breakwater is favoured, caissons would now be used, and the following remarks relate to this type of structure in particular.

Significant failure modes of a vertical breakwater are indicated in Figure 2.

The wave forces acting upon a vertical-walled structure, although not yet fully understood, are more readily quantified than those acting on a rubble mound. However, if damage occurs to a vertical-walled structure, it is usually serious involving movement of complete units, significant loss of performance and heavy reinstatement costs.

Current practice is to treat these structures by static analysis of computed disturbing and restoring forces, with an appropriate safety factor.

Having made a preliminary choice of caisson proportions, stability analyses are carried out against sliding and overturning, and checks made on bearing pressures and foundation stability.

Wave loadings arise from the largest non-breaking waves at high tide or from values of H less than H_{max}, which may break at a lower state of the tide, and calculations will be needed of all possible conditions to establish the most critical.

Several theories are available to determine the wave loading under either clapotis or breaking waves on the seaward face of the structure, and the designer will probably carry out comparative calculations to test the sensitivity to different theories.

Under certain conditions the differences in loadings can be considerable and hence to specify minimum factors of safety to be achieved may not be meaningful. Different authorities have given opinions varying from 1.0 to 2.0 as acceptable factors of safety for different conditions. A minimum factor of safety of 1.2 against the worst conditions would probably be a reasonable target. It is frequently found that sliding is more critical than overturning with a well-dimensioned section, but possible base failures need special attention and potential wave erosion in front of the structure may require heavy armour for some distance seaward.

The caisson structure is then designed for all loading conditions, temporary and permanent, in a conventional manner.

There are two important factors in considering a caisson breakwater for a particular site; the foundation conditions and the availability of a site to make the caissons.

A natural rock foundation may be very expensive to prepare in the open sea. A soft foundation may need ground replacement to ensure that the potential for settlement is small, which it must be for a rigid structure, even if the caisson is founded on a rock mound as a composite structure.

Placing prefabricated caissons from land is not very common, and floating caissons are more likely to be used as plant requirements are more modest, but sheltered water is then needed for building the caissons. Remote building sites may entail long tows of the partly completed structure and rapid action to take advantage of a "weather window" then becomes more difficult.

It is important to finalise bed preparation, and to tow out, and sink, a caisson during very low wave action. Once placed and filled, however, an individual caisson is fairly secure against moderate storms during construction.

By contrast, a rubble mound structure is vulnerable to damage throughout the construction period, particularly at the advancing scar end, although construction can proceed during moderate wave action.

In many cases, however, a rubble mound will promise at the outset to be the most economical structure.

Figure 3 shows significant ways in which a rubble mound breakwater can fail under wave loading. It will be apparent that inter-action can occur, as the movement of one element could affect another.

All these modes will be given careful attention by the designer but it is the seaward armour layer which dominates his thoughts during the structural design process. The size of armour unit will affect the sizing of underlayers, the options for construction, and, hence, the width and level of the working crest of the advancing structure. Overtopping considerations, influenced by wave run-up on the armour, will affect the design of the crest and the lee-side armour.

In a fairly mild wave climate, where large rock is available, an economical design may be found using only natural rock.

Before the necessary proving of rock quality, the designer has to find a source, which has to be within economic distance of the site and with sufficient reserves to complete the work. An existing quarry may provide most of the information but otherwise geological maps may exist to localise the search.

Political, legal and planning aspects need to be considered from the start so that when the construction begins the full implications of exploiting the source will be predictable.

The most difficult matter to evaluate is the yield of large rock sizes to be expected from a given rock deposit. Joint spacing can be assessed from boreholes, but this will not be enough to allow for the effects of blasting.

A trial blast is useful in making a better assessment of the likely armour yield, but in many instances it will not be possible to organise trial blasting until the design is completed and a contractor is on the site with the necessary equipment.

In the absence of large rock, or in the more severe wave climates, the designer will probably adopt concrete armour units rather than be constrained by available rock sizes on very flat slopes.

In theory, it should be possible to prepare alternative breakwater designs with all the 20 or more concrete units now available, and compare them for economy. In practice this is not possible and the designer will usually make his selection from very few units of which he has experience, or where he has confidence in extensive experience of others. Few designers will have the resources to develop a completely new armour unit for a specific project.

Estimating the costs of making and placing a given armour layer is reasonably straightforward. What is not so easy is to make the comparison for different units with equivalent performance. Unit weight, porosity and layer thickness will vary between units, and there is no certain method of assessing equivalent performance, which has to cover such factors as run-up and overtopping, optimum slope, effects on the underlayer, and the assessment of tolerable damage and maintenance.

A comprehensive series of model tests would help to answer many of the unknowns, but there is never enough time or money to model test all the options, so that a choice has to be made on other evidence. The designer will usually find his choice falling between complex interlocking units such as dolosse, tetrapods or stabits and the cubical types, of which the fluted cube is probably now the most popular. Prototype experience is not easy to find, and other designers' applied model tests are rarely published. Research model tests are more often available but rarely cover more than a limited range of parameters.

There is thus no general guidance to be given on the choice of armour unit and the selection will be based on the designer's own experience and judgement.

He will then probably consider a number of solutions for the structure of the armour layer based on Hudson's formula, perhaps using one or two different units of different size on a number of side slopes, building up, with each alternative, a complete breakwater cross-section to suit.

The relative resistance of armour to wave impact and inter-unit forces is less for larger units and most designers at present would probably be more conservative in deciding the weight, particularly of the complex concrete units, when the size reaches about 20 tonnes.

In most situations, a uniform slope will be used on the seaward side, and size of unit will not be varied with depth, except that the main armour may stop above the sea bed in deep water.

Most armour units will exhibit reasonably well-tested run-up characteristics under given wave conditions. Still-water level is an important factor here, and the superimposed effects of surge and wave set-up on tide levels need to be assessed. In addition, the joint probability of extreme surge occurring with extreme waves will need to be evaluated but at the initial design stage it would be usual to take, say, a 100-year return period surge at mean high water spring tides with a 100-year significant wave as determining the mean run-up condition.

Waves higher than H_S at the peak of a storm will overtop at high water, and if this is assessed to be too severe for operation, the crest would be raised to a run-up level based on a higher wave. Persistence of severe storms, and tidal range, are important inter-actions to be considered.

The design of the cap of a rubble mound break-water receives little attention in the literature, but most large breakwaters now have a massive concrete superstructure.

The function of the concrete cap, sometimes with a wave wall, is to intercept run-up, to prevent armour units being pushed over the crest, and to provide access.

A wave wall as 'A' in Figure 4 suggests useful economy by giving a greater protection against overtopping than with the cap only, or more probably for a defined level of overtopping permitting a smaller structure.

However, the forces on the vertical wall can be large, with waves striking it over the top of the armour. In addition the wave action at the top of the armour will be severe, with wave reflections from the wall.

Design 'B' in Figure 4, with the armour covering the seaward face, seems better and if a reasonable berm is provided seaward of the cap, the armour layer is likely to be more stable here. It is also easier with this design to reduce the porosity of the mound under the cap, which then sits on core material.

Lee armour stability is much improved by extending the cap so that overtopping water overshoots the top of the lee slope. A downward extension of the sea-side of the cap has been shown to provide additional sliding resistance.

The design of the cap structure will follow the general principles of quasi-static analysis used for a vertical-walled breakwater. Wave forces would be treated by assessing the run-up of the largest likely individual wave and constructing a pressure diagram, cut off at the top of the parapet.

It may be noted that the presence of a full covering of armour on the seaward berm will tend to reduce wave pressure but this reduction is not usually allowed for as armour might become displaced.

Uplift pressure under the cap would probably be taken to follow a linear distribution to zero at the lee face in the absence of any better evidence.

With all cap design, it is worth considering that some overtopping acts as a safety valve, but complications of drainage arise if there is reclamation behind the breakwater.

Having done a preliminary design of the armour layer and cap, and considered a reasonable construction level, the lee-side armour and underlayers are designed and the remaining mass of the structure is usually formed from quarry-run material. Specifications for the layers will be checked against forecast quarry yield, and the layers designed, usually on empirical ratios, to act as filters between core and armour.

It is advantageous in long-term stability to keep the first underlayer as large as practicable, and this also has benefit in giving greater stability to the partly completed work before armour is placed.

The seaward end of a breakwater needs special consideration for all types of structure. In the case of a vertical-walled breakwater the head section is usually made wider.

A rubble mound breakwater also exhibits wave phase difference, and large downrush of waves on the lee-side of the roundhead.

Angled wave attack will always occur at some point and the consensus of opinion is to use a lower stability coefficient in design and thus increase the unit weight of armour, or provide a flatter slope. This is often the most difficult feature for construction planning.

Environmental Aspects

It is important to give some attention to the environmental effects of the breakwater. One important concern is over any changes in the sea bed which by scour or accretion can alter the stability of the structure or wave loads on the structure.

Other important aspects are siltation at the harbour entrance or coastal erosion. In a region with large movements of coastal zone deposits, these features can be of financial importance.

The objective for the designer is to obtain an estimate of the costs of dealing with such changes in the coastal regime by dredging, sand by-passing, beach feeding or subsidiary coastal structures. It is even possible that a longer breakwater should be provided, or the alignment changed for these reasons. A change in layout can alter the waves reaching the structure with a need to modify the structural design.

Economic Evaluation

Having completed the initial design of alternative structures and estimated the costs of construction and maintenance, the designer will determine the preferred solution as the least in total capitalised cost.

It is likely that initial capital cost only will be the determining factor in deciding between a vertical breakwater and a rubble mound, but operational matters will be taken into account. Theoretical optimisations of capital/maintenance for a large rubble mound are considered unlikely to be meaningful at this stage.

Some optimisation of the length may however be undertaken. Variations in cost with length are readily estimated but only simple diffraction calculations of wave action are likely at this stage, based on the best estimates of occurrence of wave heights, periods and directions at the harbour entrance. The resulting changes in waves at berths would be used to compute differential ship and operating costs.

The economic and financial evaluation is then carried out to ensure that the overall benefits are properly related to the costs.

The feasibility stage will finish with the preparation of an investment programme for detailed design and construction.

It must be recognised that at this stage the designer will be quite deeply committed to the structure he has put forward, and particularly its estimated cost. It is significant that no model tests have been carried out so far.

THE DETAILED DESIGN STAGE

Further Data Collection and Analysis

Once a decision is taken to proceed to detailed design and construction, further data collection will be implemented. Ground information will become more site-specific and the marine climate will be investigated in more detail. As data is added, further analysis will be carried out, and design parameters will be refined.

Further research into soil conditions will follow conventional practices, as described earlier, but a large part of the designer's attention will be focused on the further acquisition and analysis of wave data.

Concern will be directed to improving estimates of exceedence of given wave heights and periods but more complex descriptions of sea state will be sought, particularly with regard to wave spectra.

Once the offshore wave climate has been established from the data available, the wave conditions at the inshore area of interest need to be determined more precisely than hitherto.

Refraction and shoaling are the most important influences on transforming the wave field, and there are currently a number of mathematical models available, which enable changes in wave climate between the offshore and inshore location to be predicted.

Forward tracking of wave rays of various periods will indicate any tendency for wave concentration at the breakwater site, but back-tracking would be used to enable wave spectra to be transferred to the inshore points of interest.

The effects of variations in current fields on wave refraction may need to be studied. Furthermore, energy losses by friction and breaking may warrant consideration although mathematical modelling is not yet well advanced for these aspects.

Improved Design

Much of the design procedure and the practical considerations referred to in the feasibility stage, will remain applicable during the later design stages.

With the refined design parameters, the final selection of a structural solution will be made, and probably the choice made earlier will be confirmed.

The design will include refinements to the layout for the effects of wave penetration to berths and moorings.

Mathematical models of harbour response are being developed to deal with diffraction reflections and resonances and also, in some cases, to include depth changes and refraction within the harbour area.

At present, though, physical modelling would usually be used.

The techniques are well-established using regular waves or wave spectra to provide the data needed to optimise the harbour layout.

Environmental considerations of the breakwater location will be studied in further detail, using the latest numerical models of changes to the adjacent shorelines and, also, for the study of possible siltation in navigation areas.

These studies will confirm the final breakwater layout, leaving, to be completed during the remainder of the design process, the most important task of improving the structural design and studying in more detail each of the elements of the structure. Careful checking will be carried out of all the design parameter and load cases, following the sequence describe in the feasibility stage.

The armour layer of the rubble mound will continue to be a dominant consideration, and if complex concrete units are being used, the designer may contemplate reinforced armour unit if the sizes are large. Opinions vary widely on the cost-effectiveness of such measures. Conventional reinforcement is of doubtful effec and fibre reinforcement is also questionable.

The study of ground behaviour under and around the breakwater will be finalised and the structure will receive a more searching attention to practical construction methods. The designer will be developing firm views on a construction method, even though a contractor may in the event adopt another solution.

The important decision has now to be taken on whether to undertake hydraulic model stability tests. As with many other decisions in the design process, it will not be taken at an instant, but will have been gradually formulated in the designer's mind over the previous stages and in all probability a laboratory will have been selected and arrangements made.

Nevertheless the design process can move in two ways from here, with and without model testing. It is generally accepted now that the design process of important breakwater structures, particularly rubble mounds, should incorporate model testing, but it is by no means a universal practice.

If model testing is not carried out it will usually be because the designer feels confident in his knowledge of the local maritime environment, perhaps because of previous break-water experience in the locality. Alternatively the project finance may permit a generous capital investment with the breakwater representing a small part of the whole, and an element of over-design can be incorporated and minimal maintenance assumed.

It is possible that the designer's wish to do model tests may not be accepted, and in other cases the design programme may not allow time for model testing if political and commercial pressures are large.

However, even if model testing is to be done, it is still necessary to have a completed design at this stage, which will be based on available methods and experience. It is at this stage that the designer's judgement will receive its greatest challenge. He will be constrained, on the one hand by the caution needed in assessing a design wave and applying empirical formulae and computations, and on the other hand by the considerations of economy.

Model Testing for Stability

In testing for stability, numerical models are becoming available to deal with the geotechnical stability of large breakwaters and the variations in pore water pressures and flows under wave action. Nevertheless, most designers will, at the present time, use physical modelling, even though there are doubts about how this can deal with the overall stability of the rubble mound.

The time taken to build and test physical models can be critical in the design programme, and it is therefore necessary to plan carefully what the modelling is to accomplish, and how many models are to be built.

2-dimensional flume tests are used to check a typical short length of the breakwater cross-section under increasing wave attack, until the imposed waves are higher than the design wave and, preferably, further, until the section is significantly damaged.

3-dimensional basin tests may then be employed for further study of the design, with similar test procedures. In the case of some long breakwaters, it may be necessary to adopt a scale that permits testing of only critical features such as the roundhead, a change of direction, or where an angled attack is likely.

It is in the 3-dimensional tests that any problems due to wave concentrations or angled attack and consequent instability or increased overtopping can be evaluated, and also behaviour during construction can be studied.

The major laboratories now use programmable random wave generators so that structural behaviour under wave grouping can be assessed, and wave heights and periods which can give rise to resonance are reproduced. Wave records will normally give guidance on the wave spectrum to be adopted for such testing.

The length of each part of the test should be related directly to the anticipated length of storm conditions.

The designer will need to specify the model movements to be recorded to suit the design of the armour layer, and this will be determined by the strength of the individual units. Calculations of likely wave forces and inter-unit forces due to movements will be imprecise, so experience of performance in service will be the most valuable guide.

Evaluation of Model Results

The evaluation of model results is often limited to assessing whether the probability of exceedence of conditions giving rise to a defined level of model damage is acceptable. In the majority of cases concern is only with the initiation of damage by movements in the armour layer.

Those who have witnessed model tests will know that some movements will occur at quite modest wave heights so that an economic answer is unlikely to be found just by seeking an absolute no-movement model result.

The most usual procedure has been to accept that the model correctly measures damage, and if the tolerable movements for the particular armour unit during the design storm are only exceeded by a few percent of the number of units, then the design is deemed satisfactory.

The 100-year wave was mentioned earlier as a commonly used figure to represent the design storm condition, but the probability of this being exceeded over a reasonable service life can be significant. The evaluation of model test results for larger storms is therefore necessary, so that there is some assurance that movements do not escalate rapidly. If it is required to reduce the probability of damage, a longer return period design wave would be used.

If the designer concludes that the model exhibits too much movement, the design would be changed. In practice, it may not always be possible to carry out further model testing to include all the modifications which may have been decided from the model programme, but ideally the final design should be tested.

It is in the use of improved probability assessments, not only of wave occurrence but in the whole question of structural behaviour, that advances are being made. Higher levels of probability approach are referred to in the paper on Risk Analysis, but are not yet common procedure.

After evaluations of the model results and the consideration of damage probabilities, the capital and probable maintenance costs of the acceptable solution would be calculated. Other solutions with a lesser risk could at this stage be checked to test whether an increased capital cost may result in a greater saving in probable maintenance.

In many cases, however, the evaluation of tests and the final decision to proceed with the structure as tested or modified, will be taken without such further refinement. The design will then be drawn up and contract documents prepared.

CONSTRUCTION STAGE

When construction starts, the design engineer will receive reports on how the work is conforming to the specification. Any deviations need to be reviewed for their effect on the design assumptions and, if necessary, changes made, either to site practice or to specifications and drawings.

Monitoring of the behaviour of the breakwater, particularly settlements and movements, will start during construction, and at the end of the work the designer will furnish the owner with as-constructed drawings and a monitoring specification. This latter will recommend periodic inspection above and below water level and routine line and level checks, particularly following severe storms, so that any deterioration in the structure can be attended to.

The designer should also advise the owner of the actions required for maintenance, when they should be taken, and suggest methods to be employed. The policy must be to restore the structure to what was designed and tested, either with the owner's readily available equipment or by mobilising equipment as required.

CONCLUSIONS

In presenting this brief commentary on the present design process I am conscious that other designers may take different views of what is important and would have given different emphasis to the matters mentioned.

Nevertheless, I hope my comments have served to indicate the amount of data to be collected and the degree of judgement needed in applying it to a design. The facilities for data collection, analysis and modelling which can be deployed by a designer are very extensive and increasing with time. I would however suggest that particular regard is given to the stages at which the data and techniques can be brought into the design process.

Those who are to provide the designer with improved methods have much to do, and it is to be hoped that the present initiatives in improved model testing, and in the study of prototype behaviour, will have valuable contributions to make to the design process of rubble mound breakwaters.

REFERENCES

1. U.S. ARMY COASTAL ENGINEERING RESEARCH CENTER. Shore Protection Manual. U.S. Government Printing Office, Washington DC, 1977.

2. PERMANENT INTERNATIONAL ASSOCIATION OF NAVIGATION CONGRESSES (PIANC). Report of the International Commission for the Study of Waves. Excerpt from PIANC Bulletin No. 15 (Vol II/1973).

3. PIANC. Final Report of the International Commission for the Study of Waves. Annex to PIANC Bulletin No. 25 (Vol III/1976).

4. PIANC. Final Report of the 3rd International Commission for the Study of Waves. Supplement to PIANC Bulletin No. 36 (Vol II/1980).

5   Foundation problems

W. R. THORPE, BSc, FICE, Sir William Halcrow & Partners

The two main considerations in the design and construction of breakwater foundations are overall geotechnical stability and the protection of the foundations from the adverse effects of erosion due to the action of waves and currents.

The development of successful solutions to foundation problems depends on thorough site investigation supplemented, where necessary, by hydraulic modelling.

Instrumental observation of piezometric pressures has been used successfully to monitor performance and to verify design assumptions for foundations on soft clays.

In preparing the contract documents for breakwater foundations it is important to strive for clarity and to specify realistic tolerances and criteria consistent with smooth construction procedures.

INTRODUCTION

1. The primary consideration in the design of breakwater foundations is geotechnical stability. The object of foundation engineering is the successful transfer of the weight of the structure together with static and cyclic loads to the underlying soils. The complete interdependence of the foundation and the breakwater is implicit throughout the discussion of breakwater foundation problems.

2. All foreseeable variations in the character of the subsoils must be considered in the stability analyses. Notwithstanding the differing properties of the soil strata and their arrangement, simplicity in the form of foundations is vitally important.

3. It is essential to consider the possibility of erosion of the sea bed and its effects on the stability of the breakwater. Changes in the direction and velocity of currents are a customary sequel to the construction of the breakwater and associated harbour works. Although reliable predictions of the effects of these changes on local morphology have not been developed, it appears that hydraulic modelling is an effective method of studying the phenomenon.

4. Breakwater foundation problems are reflections of the wider influence of the marine environment. As the construction of breakwaters moves into new and relatively unfamiliar environments as, for example, the permafrost and seasonal ice attack of the far North, the foundation problems will differ from those experienced hitherto.

5. In the investigation and solution of all breakwater foundation problems, the skills of the geologist, the geotechnical engineer and the coastal hydraulics engineer must engage in a multi-disciplinary exercise.

6. The purpose of this paper is to examine the breakwater foundation problems which are commonly encountered and to review the procedures for solving these problems economically.

ACQUISITION OF DATA

7. Foundation construction problems and their implications for the cost and rate of progress of the works will influence the evaluation of alternative breakwater layouts. Navigational safety and effective shelter of the harbour cannot be sacrificed simply to achieve a modest saving in the cost of the breakwater foundations or a reduction in the construction period. However, the scale of potential savings can only be assessed in the light of comprehensive and reliable soils and marine environmental data which would enable an effective and economic design to be prepared.

Site Investigation

8. It is essential to have funds to carry out an investigation commensurate with the importance of the breakwater foundations. If there is evidence of a firm sea bed and consistent load bearing capacity in the subsoils, then a limited number of boreholes may suffice. It is not prudent to assume that a handful of boreholes will reveal all the varying patterns in the soil strata over a length of several kilometres. Therefore geophysical surveys and penetrometer tests must be used to supplement the borehole exploration.

9. If the foundation design is one which relies on the progressive gain in strength of soft clays then critically important judgements will

Breakwaters—design and construction. Thomas Telford Ltd, London, 1984

65

have to be made in the light of the laboratory test results. The best available sampling equipment should be used in order to obtain the required degree of accuracy. Swift transfer of the samples to the laboratory is no less important than the sampling procedure when dealing with weak soils.

10. The site investigation should also include a preliminary assessment of the susceptibility of the sea bed to erosion induced by storm waves or currents. If erosion is judged to be a potential threat, it is unlikely that the extent of the area at risk or the severity of the problem can be judged without a hydraulic model study. Further seabed sampling may be required when the area which will be affected by wave action or intensification of currents has been identified.

Fischer and Lu (ref. 1) have drawn attention to the need for a combined effort by engineers in the geotechnical, oceanographic and hydraulic fields in order to obtain a better understanding of the erodability of marine sediments under storm waves.

Materials Survey
11. An evaluation of foundation problems which will permit a true comparison of alternative forms of construction depends on knowledge of the availability and cost of materials of construction. Therefore a materials survey should proceed concurrently with the site investigation.

12. The survey will include prospective sources of quarry stone as protection for the base or the breakwater toes. Borrow areas for sand to replace weak soils should also be inspected. After a preliminary reconnaissance to identify the promising sites, a programme of simple sampling and borehole exploration will serve to determine the yield of suitable material. The possibility of environmental restraints on the winning and hauling of sand or rock must not be overlooked.

13. Similarly, the effect of any restrictions which may apply to the dumping of spoil removed during the course of foundation construction must be investigated.

14. If the construction of the foundations together with a base section of the breakwater is to proceed well in advance of completion of the structure, it is essential to seek prior agreement of the harbour authorities on all matters relating to the safety of shipping.

DESIGN
15. The design of the foundations includes the selection of the type of substructure transferring the loads to the underlying soils or rock. It also includes the analysis of the foundations together with the breakwater so as to ensure a satisfactory margin of safety against shear failure, sliding overturning and also a check of the deformations occurring at all stages of construction.

16. Two PIANC reports of the International Commission for the Study of Waves (refs. 2 and 3) should be consulted for guidance on the design of foundations. The earlier report dated 1976, excluded soil mechanics problems from its terms of reference. However, it contains calculation procedures for wave loadings on different classes of breakwater and also provides recommended values for the factors of safety for resistance to sliding and overturning. The report emphasises that its calculation methods are based on regular waves and may not be correct for irregular waves. The report also recommends hydraulic model studies for investigating vertical wall and composite breakwaters.

17. The final PIANC report dated 1980, examines foundation stability and reviews the procedures for dealing with weak soils underlying foundations. Geotechnical problems and range of recommended solutions are considered in an appendix to the report.

18. It is quite customary to find that the site investigation provides a wide range of values for the mechanical properties of the soils and also shows erratic variations in the soil profiles. The selection of the soil properties for the foundation analysis requires the exercise of judgement and an interpretation of the test results using the least favourable values and not the mean of all results.

19. There is evidence that the impact forces of breaking waves on vertical breakwaters fitted with energy absorbing devices are considerably less than the forces on plain walls. Onishi and Nagai (ref. 4) and Mogridge and Jamieson (ref. 5) describe model experiments in which significant reductions in forces were observed. Bearing in mind the possibility that the perforated or box facing unit might sustain damage locally, it would not be advisable to take advantage of the reduction in loadings in the design of the foundations.

20. Future widening and deepening of dredged channels and basins adjoining the breakwater must be taken into account in the foundation analysis. The allowance for overdredging must be generous. A further threat to stability may arise when the foundation rests on a stratum of conglomerate or "caprock" as it is often termed in the Middle East. The technique of undermining which is used in dredging this material can result in a lowering of the sea bed for a much greater distance than might be reasonably foreseen.

21. Breakwater foundations must always have an additional margin of strength at all positions where concentrations of wave energy and turbulence may be expected, namely at the extremities, at changes in alignment and also close to a bar.

Erosion
22. Erosion of the foundations and breakwater toes is a major problem for which reliable methods of diagnosis have not been devised. The only dependable remedy is to provide effective

protection against erosion over all areas of
sea bed which are vulnerable to attack. Rubble
mound, vertical wall and composite breakwaters
are all at risk from erosive forces.

It is to be hoped that a basis for empirical
methods for dealing with erosion will emerge as
further examples of successful repairs are published in the literature of breakwaters.

23. Erosion may be induced by waves in particular the standing wave or clapotis at a vertical
wall. Sainflou showed that the maximum particle
velocities at the sea bed occur at a distance of
one quarter of the wave length from the wall.

The down rush of waves at the toe of a rubble
mound breakwater may also result in significant
erosion.

24. The construction of a breakwater will influence the currents associated with tidal streams
and when these are strong and the sea bed is
erodible, severe scour may occur. The development of a scour hole some distance ahead of the
breakwater during construction is one example
of the phenomenon. Hydraulic models, both mathematical and physical, have been used to determine the extent of the erosion and to test
methods of combating it. Olliver (ref. 6) gives
examples of the economies resulting from the use
of mathematical and physical models to the solution of individual problems. He also states
that investigations of the movement of material
in mobile bed physical models is time consuming
because of the need to prove the model for both
water and sea bed material. It is not surprising therefore that hydraulic model investigations
of breakwater foundations are seldom undertaken
and then only in the case of major habour developments.

25. A valuable case history has been provided
by Kerckaert and others (ref. 7) in their description of the planned sequence of hydraulic
model studies which resulted in the development
of successful counter erosion measures prior to
the construction of the new outer harbour breakwaters for Zeebrugge.

26. A fixed bed tidal model of the harbour ex-
tensions and the coastal region bounded by the
Scheldt estuary was used to study the changing
tidal flow patterns anticipated during the progress of breakwater construction. This model
showed that increases in the peak current velocities over an area including the entrance and
the adjoining lengths of the main breakwaters
would cause severe erosion. Approximate assessments of erosion were determined for different
stages of the progress of the main breakwaters
indicated that the maximum deepening in the
path of the breakwater would reach as much as
5m. Fig. 1 shows diagrammatically the erosion
expected to occur at an intermediate stage of
breakwater progress. In the conditions which
were foreseen the quantities of materials
needed for the breakwaters would have been
increased and the swift currents would have
added to the problems of foundation construction.

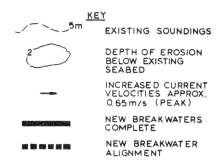

KEY

—⁀—‿—·5m   EXISTING SOUNDINGS

2   DEPTH OF EROSION
BELOW EXISTING
SEABED

→   INCREASED CURRENT
VELOCITIES APPROX.
0.65 m/s (PEAK)

▬▬▬▬   NEW BREAKWATERS
COMPLETE

■ ■ ■ ■ ■   NEW BREAKWATER
ALIGNMENT

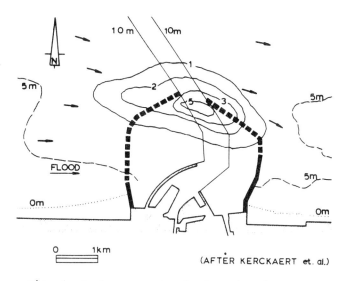

(AFTER KERCKAERT et. al.)

Fig. 1 Zeebrugge outer harbour breakwaters
Predicted erosion during construction

In the next stage of physical modelling, a
mobile bed of polypropylene grains was used in
the area subjected to increased current velocities. A sedimentological model provided
further confirmation of the loss of sea bed
material.

27. The authors concluded from the experimental studies that breakwater foundation construction should be expedited so that it would
be completed in the critical lengths prior to
the stage at which current velocities would
lower the sea bed. Since it was proposed to
build the breakwater on a sand replacement
mattress, it was necessary to place the gravel
filter layer and toe protection to prevent
erosion of the sand.

28. The full protection system was tested in
a two dimensional model in order to assess the
loss of gravel. As an additional check, the
empirically based Bijker formula was used to
calculate gravel transport. The observations
of loss of gravel in the field agreed with the
forecasts.

29. The successful execution of the outer
harbour breakwater foundations at Zeebrugge is
a striking demonstration of the contribution

of hydraulic modelling to foundation design.

SAND REPLACEMENT FOUNDATIONS

30. The difficulties of foundation design and construction increase when there is a soft bottom and unfavourable subsoils. Weak and compressible soils occur in the shore deposits on many breakwater sites. The possible solutions to the foundation problem include complete replacement of soil down to a load bearing stratum, stabilization of the soils by various methods, and spread foundations which prevent shear failure and erosion whilst allowing some measure of settlement. In principle, the choice of rubble mound construction which possesses greater tolerance to settlements without loss of serviceability, simplifies the foundation problems in poor soils.

31. Fig. 2 shows diagrammatically a sand replacement mattress which distributes the breakwater loads, and gives a satisfactory factor of safety against rotational slip. In the case of exceptionally weak soils, it will be necessary to increase the width of the replacement mattress and to provide berms in gravel or rock to act as a counterweight to increase the resistance to shear failure. The thickness and proportions of the sand replacement mattress will vary in proportion to the depth of water and the characteristics of the soils. As the foundation loads are applied, the underlying soils will be consolidated and gain strength. Therefore it is important to select the foundation material so that it will allow free drainage of the subsoils, and will also stop migration of the breakwater core.

32. It is nearly 50 years since Barberis (ref. 8) described the construction of a sand replacement foundation for a rubble mound breakwater at La Spezia. He had skilfully overcome a difficult foundation problem since the bottom consisted of at least 20m of soft clay with minimal bearing capacity. Unsuccessful attempts by Barberis' predecessors to build a breakwater by dumping stone directly on the sea bed had resulted in settlements eventually reaching 18 m, accompanied by heave, as shown diagrammatically in Fig. 3.

The sand in the replacement mattress was selected with a grain size of 0.2 mm to 0.4 mm; the thickness of the mattress varied from a minimum of 2 m to a maximum of between 5 and 6 m at the roundhead. During the three year construction period settlements of 0.5 m were observed. A further settlement of 0.3 m took place after another 12 years; thereafter settlement had practically ceased.

33. One of the principles of the sand replacement foundation is the progressive application of the load as the strength of the subsoil increases due to consolidation. Consequently, the more accurate the soils data, the better the engineer will be able to analyse the foundation and plan the timetable for the successive increments in the height of the breakwater.

Measurement of Performance

34. It is sound practice in foundation engineering generally to monitor performance. Simple level determinations can be carried out on breakwater foundations and the observation of settlement should be regarded as an indispensable feature irrespective of the existence of foundation problems.

35. Direct levelling points have to be located at positions which will not be disturbed during breakwater construction. Even if they do escape harm during construction, there are self evident problems in retaining the monitoring positions

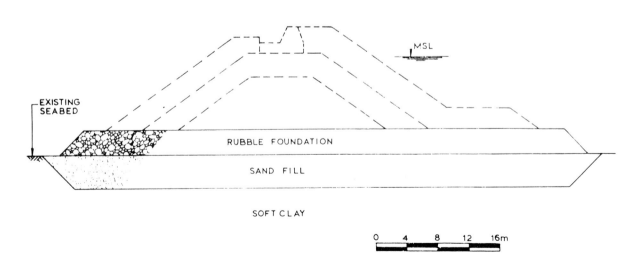

Fig. 2 Typical Cross Section of Sand Replacement Foundation

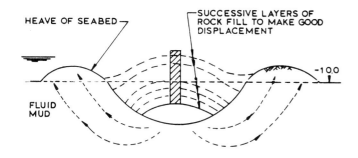

a) BREAKWATER WITHOUT SAND REPLACEMENT FOUNDATION

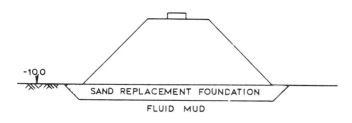

b) BREAKWATER WITH SAND REPLACEMENT FOUNDATION

Fig. 3 La Spezia Breakwater (after Barberis)

in open sea conditions. In any case, the readings will not provide direct evidence of the behaviour of zones of primary interest.

36. Geotechnical instruments which are sufficiently robust to be installed in the foundations and underlying soils are now available. The instruments will record automatically soil and pore water pressures and ground settlement. It is also possible to install a series of inclinometers in a single column to give indications of transverse movement. The considerable outlay on geotechnical instrumentation, even for a few observation points, combined with practical difficulties of installation militate against the provision of the equipment unless the need to compare performance with the design assumptions is critically important.

37. An instrumentation scheme of outstanding importance at Cagliari has been described by Albert and others (ref. 9) and the further reports on performance, which are to be published in future PIANC bulletins will be a valuable addition to the literature of foundation engineering. The soils underlying the rubble mound breakwaters consist of weak muds, silts and clays extending to a depth of not less than 20m.

38. Instrumented sections have been established

at selected positions along the breakwaters. Ground settlement sensors and earth pressure cells have been placed on the sand replacement foundation. Piezometers and fixed gauge inclinometers have been installed in the subsoils. The authors report that analysis of the data recorded by the instruments has permitted reductions in the originally planned waiting periods between successive increments in the height of the breakwaters. The authors also emphasise the exceptional care needed to carry out the installation successfully.

39. Problems of foundation settlement must be identified during the planning and design of the works so that appropriate measure to contain settlement can be evaluated before construction. The cost of trying to make good foundation settlements after they have occurred will be prohibitive.

40. Differential settlements of the foundations of a rubble mound breakwater need not result in loss of serviceability. When settlements are anticipated, it is desirable to construct the foundations and base layer of a rubble mound breakwater in advance so that the deformations take place prior to breakwater construction.

41. If the foundations beneath the toes of rubble mound or sand asphalt breakwaters yield sufficiently to allow the armour layer and core to separate the structure will be extremely vulnerable.

42. Simoen, Vandenbossche and Brouns (ref. 10) describe the action taken to densify the replacement sand and underlying soils beneath the toes of a sand asphalt breakwater within the new Zeebrugge development. The remedy consisted of vibrocompaction which was carried out successfully using plant mounted on a jack-up platform.

43. One case history dealing with the use of dynamic consolidation to improve the resistance to liquefaction of a breakwater foundation has been published. Gambin (ref. 11) describes the successful treatment of very loose silty sand at Kuwait. A 1.5 m thick rockfill blanket was dumped on the sea bed prior to consolidation. The pounder was 32 tonne weight and it was dropped between 3 and 10 times, firstly on a 4 m grid and finally on a 2.5 m grid. A significant improvement in the density of the very loose silty sand was recorded. The operation appears to have been carried out in approximately 10 m depth of water.

44. Gravity vertical wall and composite breakwaters do not tolerate uneven foundation settlement. If liquefaction or undue deformation of the subsoils is anticipated, they must be replaced or improved in depth.

45. Large scale replacement introduces operations which must follow a carefully planned sequence. To illustrate this further, the dredging passes must avoid disturbance of the remaining soil and should also minimise sil-

tation. Filling must proceed without delay. The disposal of the spoil must comply with environmental regulations.

46. Each of the foregoing activities will have to comply with the corresponding tolerances and acceptance criteria. It follows that clarity and simplicity in specifying the control testing is of the greatest importance to the smooth running of the replacement procedures.

47. In principle it should be economically advantageous to leave the unsatisfactory soils in place and improve them by deep densification using purpose built and robust equipment designed for open sea conditions.

48. Experimental work associated with the Oosterschelde Storm Surge Barrier in The Netherlands has resulted in the development of techniques which might be applied to breakwater foundations. Pladet (ref. 12) describes a powerful top mounted vibrator which creates a column of densified soil by means of radial fins attached to a steel tube. In its field trials, the equipment was effective in loose cohesionless soils provided the silt content did not exceed the range 10% to 15%.

CONCLUDING REMARKS

49. The influence of breakwater foundations on methods of construction, progress of the works and capital and maintenance costs is not in question. Recent case histories dealing with foundation practice demonstrate the importance of a multi-disciplinary approach in the identification and solution of foundation problems. It is to be hoped that these case histories will prompt similar accounts which will benefit everyone engaged in maritime civil engineering.

REFERENCES

1. FISCHER J.A. and LU T.D. Storm wave induced erosion of marine soils. Fifth Symposium of the Waterway, Port Coastal and Ocean Division of the ASCE "Coastal Sediments 77", pp. 885-897.
2. PIANC. Final Report of the International Commission for the Study of Waves, Annex to Bulletin No.25 (Vol. III/1976).
3. PIANC. Final Report of the 3rd International Commission for the Study of Waves. Supplement to Bulletin No.36 (Vol. II/1980).
4. ONISHI H. and NAGAI S. Breakwaters and sea walls with a slitted box-type wave absorber. Proceedings of the ASCE Specialty Conference "Coastal Structures 79", pp. 9-28.
5. MOGRIDGE G.R. and JAMIESON W.W. Wave impact pressures on composite breakwaters. Proceedings of the 17th Coastal Engineering Conference 1980, pp. 1829 - 1848.
6. OLLIVER G.F. Harbours : the value of physical modelling. Proceedings of I.C.E. Conference on Hydraulic Models, 1981.
7. KERCKAERT P., LAFORCE E., ALLAERT J., NEYRINCK L., and DE ROUCK J. Planning and construction of bottom layers in sea gravel in relation to the foundations of breakwaters. Symposium Engineering in Marine Environment, Brugge 1982.
8. BARBERIS M.C. Recent examples of foundations of works resting on poor subsoil. PIANC 16th Congress, 1935. Report No. 111.
9. ALBERT L., ALMAGIA E., BARATONO E., COSTA G., GRIMALDI F., NOLI A., and TOMARELLI S. Site investigation, instrumentation and construction of harbour works at Cagliari, La Spezia, and Giola Tauro, PIANC 25th Congress 1981.
10. SIMOEN R., VANDENBOSSCHE D., and BROUNS P. Zeebrugge port extension - design and construction features of service port breakwaters. Symposium Engineering in Marine Environment, Brugge 1982.
11. GAMBIN M. Menard dynamic consolidation, a new method for improving foundation beds offshore. Symposium Engineering in Marine Environment, Brugge 1982.
12. PLADET A.A. Densification of the subsoil in field practice, results obtained with a deep compaction method. Symposium on foundation aspects of coastal structures, Delft 1978.

6   Hydraulic modelling of rubble mound breakwaters

M. W. OWEN, MA, MICE and **N. W. H. ALLSOP**, BSc, Ports and Coastal Engineering Group, Hydraulics Research Station

SUMMARY
The paper describes briefly the various steps taken to model the breakwater construction and the wave climate to which the breakwater is to be subjected. Some of the methods available to assess the hydraulic performance and stability of the breakwater are discussed, and a typical model study programme is outlined. Finally, the possibility of model 'scale effects' is examined.

1. INTRODUCTION

The primary objective of model testing of rubble mound breakwaters is to check the stability of the breakwater up to and exceeding the design sea state. However, modelling is also used to gather information on the hydraulic performance of the breakwater, in terms of reflection, run-up, over-topping and wave transmission. This information can then be used in the design process for the breakwater location, length and alignment to provide optimum wave protection for the harbour or other coastal installation. This paper deals mainly with breakwater stability, especially as applied to rubble mound breakwaters.

Rubble mound breakwaters consist essentially of a mound of quarried rubble protected by an armour layer of large rocks or concrete blocks. To prevent the fine material of the core from being leached out through the coarse armour layer by wave action then one or more filter layers of graded stone are introduced between the core and armour. Finally, a roadway is often added on top of the breakwater, so that a typical breakwater might have a cross section similar to Fig 1.

Fig. 1 Typical breakwater cross-section

For such a breakwater, and for a given wave climate, the breakwater stability (its resistance to storm damage) depends on a long list of parameters, including amongst others such items as:-

armour block shape, weight, density and strength;

interblock friction and interlocking;
armour layer packing density and permeability;
underlayer porosity and permeability;
height and slope of breakwater.

Despite extensive research over many years the importance of most of these parameters is not fully understood, and the most widely used formula for the initial design of breakwaters groups all but three of these parameters into an empirically derived damage coefficient K_D. This formula also takes no account of a wide range of wave climate parameters other than wave height. It is therefore clear that rubble mound breawaters have to be designed very largely on the basis of past performance and/or on scale model studies. It is also apparent that any model should reproduce as closely as possible details of the breakwater construction and of the wave climate to which it will be subjected.

2. MODEL BREAKWATER CONSTRUCTION

The methods generally used to model the breakwater construction will be described in turn for the core, the filter or underlayers, the armour layer and finally the crown wall or roadway.

The breakwater <u>core</u> is hydraulically the least important part, and usually has low permeability and porosity in the full-size breakwater. In the model the size grading of the core material is reproduced as closely as practicable, and the specific gravity is approximately correct. The model core material is placed by hand to the correct cross-sectional shape and is damped down to achieve good compaction. The main function of the <u>underlayers</u> is to act as a filter, and for this purpose the size and shape of the underlayer rock are critical parameters, since they define the size grading of the interstices. The underlayer must also support the armour layer units through friction and interlocking, so the surface texture of the underlayer is important. The size and shape of the underlayer rock is

Breakwaters—design and construction. Thomas Telford Ltd, London, 1984

71

geometrically modelled according to the basic model scale λ, and the rock placed by hand according to the required cross-section. The porosity of the underlayer is checked by comparing the measured packing density of rock against the required density. However, the hydraulic performance of the underlayer also depends on its permeability, which is a function both of the interstices and also of the viscosity of the water. As long as water is used as the model fluid the viscosity scale is approximately unity, and the model permeability cannot be scaled correctly. This point will be returned to later, in the discussion on scale effects.

The armour layer is probably the most important part of the breakwater, and the modelling problems involve reproduction of the armour units, and of the methods and patterns of placing them. The armour units themselves should ideally have the correct shape, size, specific gravity, surface roughness, and strength. The shape and surface roughness are normally modelled geometrically according to the basic model scale λ. However, no suitable material has yet been found which will simultaneously model both the specific gravity and the strength of concrete or rock (except that special and extremely expensive low-strength micro-concrete armour units could perhaps be used if the model scale was no smaller than about 1/10 of the full size). For most models the specific gravity is modelled accurately. Unless the main armour is large rock, in which case natural stone is used, the model armour units are manufactured either from concrete or from weighted plastic, Fig. 2. With weighted plastic it is essential that the weighting material is uniformly distributed throughout the armour unit otherwise the moments of inertia of the armour unit are incorrectly modelled, and any rotational motion of the armour units will be poorly reproduced. The submerged weight of the armour unit is the important factor, ([1],[2]) so that the specific gravity of the model armour units is usually a few percent lighter than full-size. Most model testing is carried out in fresh water tanks, whereas most breakwaters are situated on maritime coasts. The reduction in

specific gravity is in direct proportion to the ratio of the specific gravities of fresh and sea water respectively.

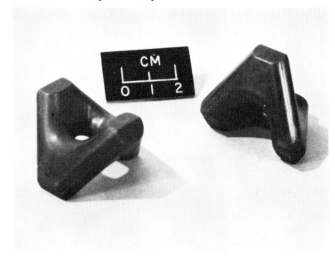

25g Stabits, Nylon with Barium Ferrite

Fig. 2 Example model armour units

The placing of the armour units should follow as closely as is practicable the methods proposed for constructing the full size breakwater. Most armour units are laid to a specified grid pattern, and sometimes to a preferred armour unit orientation. The modeller must work to the same specification, but he must resist strongly the temptation to work each armour unit into the most favourable position and alignment. His manual dexterity will far exceed that of the contractor's plant, especially under water. For this reason for most of these breakwaters built at HRS in a narrow wave flume the units are laid to the specified pattern by 'model crane'. A model swell sea may be run if the breakwater is to be constructed where calm sea conditions very rarely occur. Where a complete breakwater has to be modelled, involving several thousand armour units, Fig 3, this method of placement would be far too time consuming. The units have to be placed by hand, making sure that the correct packing density (number of armour units per given area) is achieved, but again taking care not to obtain the optimum positioning of each unit.

307g Tetrapods and 307g Antifer cubes.
Portland cement with Alag aggregate

Fig. 3 Complete model of Douglas Harbour
         breakwater

The geometric proportions of the crown wall can be correctly modelled according to the basic scale λ, and the specific gravity reproduced as described earlier. However, in the full size breakwater the crown wall is usually constructed in lengths cast 'in-situ'. This is impractical in the model, and the wall is precast in the correct lengths, and then placed on the breakwater. This may mean that the crown wall is less well keyed to the underlayers in the model. The implications of this difference on the stability of the model crown wall will be discussed later under scale effect.

For breakwaters which are expected to overtop the same general ideas discussed above apply also to the landward face, taking most care with the modelling of any rocks or armour units which are directly exposed to wave action.

3. MODELLING WAVE CLIMATE

For a given breakwater at a particular location the design engineer has at some stage to select a design storm which then has to be reproduced in the model. There is still considerable discussion about which parameters are necessary to fully describe the storm. The classic Hudson equation ([3]):

$$W = \frac{\gamma_s H^3}{K_D (S_r - 1)^3 \cot \theta}$$

includes only the wave height H, all other parameters being incorporated into the empirical damage coefficient K_D. However, the formula was derived for regular wave tests, whereas natural seas have a wide range of wave height. In applying the formula to random waves it is not immediately obvious whether H should refer to maximum wave height, mean wave height or to significant wave height (which is the normal interpretation). Since the weight of armour unit required to resist movement is proportional to the cube power of wave height, then in a random train of waves reproduction of the magnitude and frequency of occurrence of the highest waves is very important. It has also been observed that large waves tend to arrive in groups, and that a group of moderate waves sometimes causes more damage than a single large wave.([4]) It is therefore important that the form and frequency of the wave groups are reproduced. For rock armoured breakwaters all the test results have shown that wave period is not important (hence its omission from the Hudson equation). For some concrete armour units it is now thought however that wave period is important.([5]) Reproduction of the amount of wave energy present at each wave period (the energy-frequency spectrum) is therefore essential. Similarly the correct reproduction of the wave directional energy spectrum was felt to be important for some armour units. However, recent tests carried out at HRS using dolosse units (thought to be among the most sensitive to wave direction) showed very little difference in damage rates between unidirectional and multidirectional waves([6]). Finally the form of the waves as they strike the breakwater has to be reproduced if possible, especially if waves

are breaking at or near the structure.

The methods of modelling the wave climate to best meet all these requirements differ considerably from laboratory to laboratory. Apart from differing levels of sophistication, there are also divergent opinions about the causes of the occurrence of wave groups. The wave generators used at HRS give pseudo-random waves having the required wave energy-frequency spectrum. The waves are pseudo-random to the extent that a sequence of waves can be generated having random distributions of wave heights, periods and phases, but the particular wave sequence can be repeated. One particular attribute of these generators is that the sequence length (or repeat cycle) can be varied from a few seconds up to an infinite time, without in any way affecting the wave energy-frequency spectrum. This facility allows the use of very long sequences (typically many hours in the model) when testing breakwaters, sea-walls, or harbour disturbance. With very long sequences of random waves it can be shown that wave height (in deep water) follows the classic Rayleigh distribution, Fig. 4, and with random phases wave groups are automatically generated of the correct form and frequency of occurrence. The fact that the wave sequence is repeatable also means that comparative tests can be run on different breakwater designs using exactly the same sequence of waves.

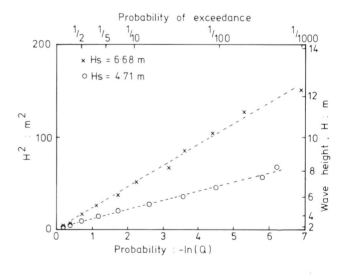

Fig. 4   Rayleigh distribution of wave heights in random wave flume

Other laboratories advocate the use of actual wave sequences recorded at the site for model testing of breakwaters. The argument is that the physics of wave groups is incompletely understood, and it is better therefore to use recorded wave trains rather than attempt to generate them in the laboratory. However, this method suffers from major disadvantages. Firstly, very long records are required to obtain an accurate description of the frequency of occurrence of extreme wave heights and wave groups for a particular sea state. If any past records are available (which is in itself unusual) they are almost certainly no longer than about 20 minutes each. Reproduction and repetition of a 20 minute record is in no way representative of the full wave height/wave groups statistics of the sea state. Secondly, it is most unlikely that wave records are available for a storm even approaching the magnitude of the design storm. The records for the most severe storm experienced have therefore to be extrapolated to represent the design storm. This extrapolation is achieved by amplifying the heights of the waves in the record and stretching the time base to achieve the required design significant wave height, and mean wave period. The occurrence of wave groups within the record is therefore unchanged: however if the physics of wave groups is not known (as is claimed) there can be no justification for assuming that the 'groupiness' of the waves during the design storm is equal to that during the worst storm recorded. Some extrapolation is of course necessary in the method of wave generation used at HRS and at some other laboratories, but in this case the extrapolation is based on predicting the wave energy – frequency spectrum for the design storm, a much more fully understood parameter.

Finally, most breakwaters are constructed in relatively shallow water, where the design wave is probably limited by breaking, and where the spread of wave directions is greatly reduced by refraction. In this case the normal practice in laboratories of using uni-directional waves is probably justified. Occasionally, however, breakwaters are built in deep water, where waves have a wide spectrum of directions, and for these situations model testing ought properly to be carried out using multi-directional random waves.

4. TEST SEQUENCE AND DAMAGE ASSESSMENT
Although a breakwater is designed for a given wave condition, such a condition cannot occur abruptly – it simply represents the peak situation in the gradual build-up and decay of the design storm. When testing a specific breakwater the test program should therefore begin with wave heights of about 50% of the design wave, and increase gradually in steps up to the design wave height. If the breakwater is still more or less intact at this stage then wave heights should be increased further up to about 20% higher than the design value. This is necessary to ensure that complete stability of the breakwater at the design wave condition does not mean that the armour has been significantly overdesigned. It is also a reflection of the fact that the derivation of the design wave condition is usually with inadequate information, and cannot be assumed to be fully accurate.

The full test sequence therefore involves wave heights increasing in steps from about 50% to 120% of the design wave condition. The duration or number of waves to be used at each wave height depends very largely on the amount of information available. If virtually no information is available, then each wave height is usually run for a fixed number of waves, say 1000 or perhaps 5000. As the wave period will

be changing with wave height (since the offshore wave steepness stays approximately constant during the build-up of a storm) the actual run times for each part of the test will differ. If a reasonable amount of information is available, then it might be possible to synthesise the time history of the design storm. For example, for a recent project carried out at HRS many years wave data was available. By examining all the major storms within that time (the largest of which was about 70% of the design wave) it was possible to derive the design storm profile shown in Fig 5.

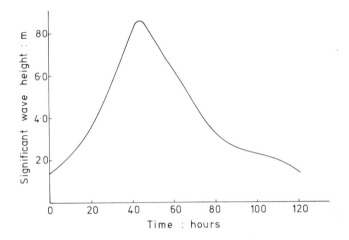

Fig. 5 Example of design storm profile

This figure was then used to select the duration of each wave height during the test sequence. For example, wave heights between 4.5 and 5.5m were expected to occur in total for about 10.6 hours during the storm. In the test sequence this was represented by a wave height of 5.0m for 10.6 hours. For wave heights above the design peak value the minimum duration of three hours was used.

Two main advantages arise from testing with gradually increasing wave heights rather than starting immediately near the design value. Firstly, if the armour happens to be under-designed the wave height at failure will be determined. The breakwater can then be quickly redesigned using this information. Secondly, it is more representative of the actual conditions, with the lower wave heights tending to settle down the breakwater before the attack by the largest waves. During each part of the test at a constant sea state the behaviour of the breakwater is closely observed and damage is assessed. At HRS the breakwater is very carefully photographed at the beginning and end of each part of the test. Camera positions are rigidly located perpendicular to the face(s) of the breakwater. The photographs are then printed on translucent paper to exactly the same scale. By superimposing prints of successive parts of the test the aggregate movement of armour units occurring between start and finish of each part of the test can be easily assessed. Alternatively, each print can be compared with the photograph taken immediately prior to commencement of testing. This will then give the accumulated damage through the tests up to that point.

Traditionally, the damage to rubble-mound breakwaters has been categorised in terms of the number of armour units which have been totally dislodged from the armour layer. This number may be expressed as a percentage of the total number of units on the armour face. However, this method assumes that the armour units remain intact at all times. With the original rock-armoured breakwaters this was probably a reasonable assumption. With some of the more slender concrete armour units it is known that breakage can occur at wave heights lower than those required to pluck armour units off the face of the breakwater. For these units the ideal model testing would reproduce this breakage: however, as mentioned earlier this is not generally possible. Instead very much more attention is now being paid to the movements of the model armour units. For example, for a recent project at HRS four categories of armour unit movements were used:

0 - no discernible movement
R - unit seen to be rocking, but not permanently displaced
1 - unit displaced by up to 0.5 D
2 - unit displaced between 0.5 and 1.0 D
3 - unit displaced by more than 1.0 D

D is a typical dimension of the armour unit (possibly its height or equivalent cube size $^3\sqrt{(W/\gamma_s)}$ and category 3 corresponds closely with the traditional 'extraction' of the unit from the armour layer. Category R can be determined visually either by observation during the test, or by analysis of video records, or of single-shot cine film. The other damage categories can be determined easily from the overlaid photographs. The number of units falling into each category is counted, and the result expressed as a percentage of the number of units present in the armour layer, Fig. 6.

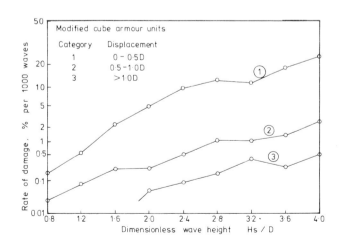

Fig. 6 Example of armour layer damage results

Unfortunately, there is not as yet any uniform standard for categorising the damage in this way, and at present the choice of movement

limits for each category is arrived at by discussions between the design engineer, the modeller and ideally the concrete technologist. These limits depend on the amount of movement which a particular type of armour unit is expected to be able to withstand before fracturing. The example categories listed above were for tetrapod and modified cube armour units. Different categories might well be necessary for more slender units. Clearly, there is a lot of scope for further research into the amount of movement which occurs on breakwaters, and the amount of movement which can be tolerated by an armour unit before breakage occurs.

As well as checking the stability of the design the modeller must also determine the general hydraulic performance and the effects that the breakwater will have upon the surrounding area. Waves reflected from a structure may increase beach erosion in front of the structure, and/or may worsen sea conditions for navigation. It is, therefore, important to determine the reflection characteristics of the design. Similarly, the wave run-up and draw-down performance must be determined in order to set the upward and downward extent of the armour layer. Under very extreme conditions, some waves may overtop the breakwater, exciting waves in its lee. Further, as the breakwater is essentially permeable, some wave activity may be transmitted through it. Overtopping and transmission will increase the wave activity behind the breakwater and must be allowed for by the designer.

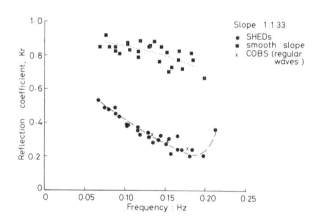

Fig. 7 Reflection coefficients for smooth and hollow cube slopes

To determine the reflection characteristics, waves are measured in front of the breakwater using two twin wire wave probes. These measurements are analysed by the method of Kajima[7] to yield the incident spectrum, reflected spectrum and hence the reflection coefficients, Fig. 7. These coefficients, Kr, are defined $Kr = \sqrt{Sr/Si}$, where Sr and Si are the reflected and incident energy in a frequency band width and Kr is the reflection coefficient in that band width. Run-up and draw-down over, on or in the armour layer may be measured with special gauges, if it is possible to locate such a gauge at the requisite level. In a recent study on a regularly placed armour block, the run-up performance was assessed from video records by counting waves exceeding predetermined levels. This method allowed easy comparison with the Rayleigh probability distribution, Fig. 8.

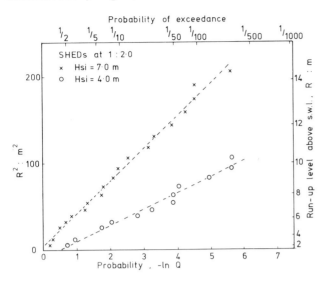

Fig. 8 Rayleigh distribution of wave run-up over a hollow cube armoured slope

Wave transmission through the breakwater may be measured simply by interposing a wave probe between the model breakwater and the energy absorbing beach at the end of the flume. The transmission coefficient Kt is usually defined in the same fashion as Kr. Transmission of wave energy by overtopping will also be measured by that probe. It is useful also to record the number (or percentage) of waves overtopping the structure. A variety of different gauges and sensors have been used successfully in recent projects. The overtopping discharge may also be measured quantitatively using special collection tanks.

5. TYPICAL MODEL STUDY

Model studies of breakwater stability are not cheap, because they require fairly expensive equipment, such as sophisticated wave generators, and they are at the same time fairly labour intensive. Despite the fact that failure of a full-size breakwater would cost many times more than the model study, it is nevertheless important that the model studies proceed in a way which ensures best value for money. This ideal is most likely to be achieved if the design engineer and the modeller work closely together. This involvement should start at the very beginning of the project with discussions about the preliminary design. The experienced modeller tests several different breakwater designs during a year, and can often offer advice on particular aspects of the preliminary design before model tests begin. Most of the obvious weak points can therefore be eliminated beforehand.

The actual model studies will typically fall into four phases - outline tests in a wave flume or channel, detailed tests in a wave flume, detailed tests in a wave basin or tank, and finally construction stage tests. The outline tests on the proposed breakwater cross-section should be as brief as possible, preferably with

the design engineer present. The exact form of these tests will vary from project to project, but the main aim should be to obtain a rough estimate of the performance of the structure at the design wave condition, and to identify critical parts of the design 'cross-section'. Various modifications can be examined, to determine which is the most promising design. Once this has been achieved, the preferred cross-section should be subjected to much more detailed testing, Fig. 9. Each of the tests during this phase of the model study should follow closely the procedures described in parts 2, 3 and 4 of this paper. This means that the model break-water should be completely reconstructed between tests. Each stability test should span a range of wave heights between about 50 and 120% of the design wave. Damage analysis should incorporate determination of armour unit rocking and displacement. Each test should be repeated at least once and the range of tests should incorporate, where appropriate, variations in such parameters as water level, wave steepness, wave spectral type, etc.

Fig. 9   Flume test for Douglas Harbour break-water

When the design of the breakwater cross-section has been finalised testing of the breakwater should be transferred to a wave basin for examination of the performance of the breakwater roundhead and/or the performance of the breakwater trunk in angled wave attack. The same approach should be adopted as for the flume studies namely a few brief tests followed by more detailed testing. The brief tests would identify possible trouble spots in the design, and also determine the critical wave direction. The detailed tests would then examine in detail the most promising design.

Finally, construction stage tests should be carried out to assess possible damage to the structure during construction. The basic difficulty is that construction of the break-water core must necessarily advance ahead of the construction of the underlayers, and of the underlayers ahead of the armour layer. At least three stages of construction therefore exist simultaneously along the breakwater. The problem is therefore essentially three-dimensional and the construction phase tests should be carried out in a wave basin. Ideally, these tests should be performed after the contractor has been appointed because the contractor and the design engineer often differ on the exact sequence and method of construction. The tests should aim not only to establish the extent of any damage after a storm, but also to determine the final location of all the materials forming the breakwater, so that an assessment can be made of the possibility of recovering some of the material. For the new breakwater at present nearing completion at Douglas, Isle of Man, both wave flume (Fig. 9) and wave basin tests (Fig. 3) were carried out for the design engineer (Sir William Halcrow and Partners) and these were followed by wave basin tests on four critical stages in the proposed construction programme, Fig. 10, carried out for the contractors (French Kier Construction Limited).

Fig. 10 Douglas Harbour breakwater. Anticipated construction stage Summer 1982 prior to model testing

6. SCALE EFFECTS
In the modelling of open boundary hydraulics, phenomena which are primarily dependent on gravitational and inertial effects are modelled according to Froudian laws. Thus for the reproduction of gravity waves the wave heights are modelled according to the basic scale λ, and the wave periods are modelled as $\lambda^{\frac{1}{2}}$. However, when water is used as the fluid both in proto-type and model then secondary forces may be exaggerated, because for example surface tension and viscosity are not correctly scaled. 'Scale effects' arise when these secondary forces become significant. For example, it is known that water waves with periods less than about 0.5s cease to become gravity waves and are governed instead by surface tension. If ocean waves are modelled at such a scale that the model waves have periods less than 0.5s then scale effects will arise - the model waves will not behave in the same way as the ocean waves. For breakwater modelling the lift and drag on individual armour units are dependent on the Reynolds number, which is a measure of the relative importance of inertia and viscous forces. At the low Reynolds number of small laboratory models the lift and drag forces may be relatively larger than is the case under strict dynamic similarity. If significant this would result in the model showing too much damage and hence lead to overdesigning. Similarly, the permeability of the armour - and under-layers will depend heavily on the Reynolds number: at low values the permeablity is very

much less which again might lead to increased damage and increased overtopping on the model breakwater. Unfortunately, because a full understanding of the hydrodynamics of breakwater stability is still a considerable way off no theoretical value can yet be placed on the threshold for model Reynolds numbers to avoid scale effects. Instead, series of model experiments have to be carried out at several different scales, and the results compared. Some model tests have been carried out at various laboratories, but unfortunately the results have been somewhat contradictory. Hudson and Davidson ([8]) (using regular waves), recommend that Reynolds number based on the armour units,

$$Re = \frac{(W/\gamma_s)^{1/3}\sqrt{gH}}{\nu}$$

should always be greater than about 3.0×10^4, where H is the wave height at zero damage. They also gave a graph to enable the observed model damage to be corrected if the Reynolds number was actually smaller than this threshold. How-ever, random wave tests by Thompson and Shuttler ([9],[10]) and by Torum et al ([11]) carried out at scales including those where substantial over-damage was predicted showed no such effects. The results of all tests at different scales agreed well with each other and with recorded prototype damage. It seems likely therefore that although it is known that Reynolds number has an effect on lift, drag and permeability the accumulated effect on the stability of the breakwater is negligible. As far as the break-water armouring is concerned the selection of the appropriate scale for breakwater model tests is therefore not very critical. However, most laboratories and design engineers would probably agree that bigger models are better, but usually also more expensive.

The possibility of scale effects arising in modelling the stability of the crown wall has been mentioned earlier. Differences in methods of constructing the full-size and model crown walls may lead to an over-prediction of crown wall movement. However, the reduced permea-bility of the armour - and under-layers in the model may lead to reduced uplift and horizontal thrust on the crown wall, thus giving a tendency to under-predict crown wall movement. It is not possible to predict which of these effects dominate, or whether they cancel. Considerable further research is needed to establish design methods for crown walls, and to investigate problems associated with their modelling.

7. CONCLUSIONS
Because there is still so much unknown about the performance and stability of rubble mound break-waters, the construction of the model breakwater and the reproduction of the wave climate should follow as closely as practicable the actual pro-totype conditions. In the absence of any practical alternative, unbreakable armour units are used in the model. Because prototype units can break considerable attention must be paid in the model to the rocking and relatively minor movements of the armour units. Merely recording

the number of units extracted from the armour layer does not provide enough details of possible damage. The typical model study should ideally incorporate wave flume tests on a typical cross-section of the break-water, wave basin tests on the breakwater round-head and finally wave basin tests on the various construction phases. These last tests should preferably be undertaken after the contractor has been appointed.

8. REFERENCES

1. PRICE W A. "Static stability of rubble mound breakwater". The Dock and Harbour Authority Vol 60 No. 702, May 1979.

2. ZWAMBORN J A. and VAN NIEKERK M. "Additional model tests, dolos packing density and effect of relative block density". CSIR Research Report 554 Stellenbosch, South Africa.

3. COASTAL ENGINEERING RESEARCH CENTRE. "Shore Protection Manual". U.S. Government Printing Office, revised periodically.

4. BURCHARTH H F. "The effect of wave grouping on on-shore structure". Coastal Engineering 2 (1979) 189-199.

5. WHILLOCK A F. and PRICE W A. "Dolos blocks as breakwater armouring". Proc. 15th Coastal Engineering Conference, Honolulu, 1976.

6. SHUTTLER R M. "The effect of wave directionality on dolosse armour stability" - in preparation.

7. KAJIMA R. "Estimation of an incident wave spectrum under the influence of reflection". Coastal Engineering in Japan Vol 12, 1969.

8. HUDSON R Y and DAVIDSON D D. "Reliability of rubble-mound stability models". ASCE Symposium on Modelling Techniques, San Francisco 1975.

9. THOMPSON D M and SHUTTLER R M. Design of slope protection against wind wave attack, CIRIA Report 61, London, 1976.

10. SHUTTLER R M. Some aspects of modelling slope protectin, BNCOLD Conference, Keele University, September 1982.

11. TORUM A, MATHIESEN B and ESCUTIA R. Scale and model effects in breakwater model tests, 5th International Conference on Port and Ocean Engineering under Arctic Conditions, Trondheim, Norway, 1979.

9. NOTATION

D Equivalent size of armour unit $(W/\gamma_s)^{1/3}$

H Wave height

K_D Empirical damage coefficient (Hudson' coefficient)

K_r Reflection coefficient

K_t Transmission coefficient

S_i Incident wave spectrum

S_r Reflected wave spectrum

S_t Transmitted wave spectrum

s_r Ratio of specific gravity of armour unit to specific gravity of water

W Weight of armour unit

γ_s Weight density of armour unit

Θ Angle of breakwater slope (above horizontal)

λ Basic model scale (prototype dimension/model dimension)

γ Kinematic viscosity

10. ACKNOWLEDGEMENTS

Research on rubble mound breakwaters at the Hydraulic Research Station is sponsored jointly by the Departments of the Environment and of Transport and the Ministry of Agriculture, Fisheries and Food. The paper is published by permission of the Director of Hydraulics Research.

Discussion on Papers 5 and 6

J. QUINLAN *(Brian Colquhoun & Partners)*
Mr Thorpe refers to a paper by Gambin (ref. 11) concerning the densification of foundation beds offshore. I would like to provide a few details on the development and execution of the work.

Dynamic consolidation of the sea bed beneath a breakwater. In the construction of a breakwater in the Arabian Gulf, dynamic consolidation of loose soil beneath the sea bed had to be used as a result of the acceptance of an alternative tender for the breakwater design. The alternative design used patented, voided concrete blocks. The voids dissipated a proportion of the wave energy, but nevertheless substantial horizontal forces were imparted to the structure, giving significantly higher bearing pressures than under the original rubble-mound design.

The soil beneath the breakwater (Fig. 1) comprised a thin sandstone caprock overlying loose to very loose, silty, organic, carbonate sands (with silt content up to 35%) extending to a depth of 4 or 5 m before becoming denser and less silty. Dense siliceous sands occur at a depth of about 8 m. Standard Penetration Test (SPT) results of $N = 1$ to 3 blows were typical of the very loose deposits, and when the SPT tool was driven continuously it sank nearly 2 m under its own weight as soon as it entered the very silty sand. This suggested that liquefaction had occurred, and when an intact sample of the very silty sand was recovered using a Bishop Sampler and placed upon a tray, collapse and flow of the sample occurred after a few light blows to the tray.

Three problems were posed in the design of the foundation: to analyse reliably the bearing capacity; to eliminate the liquefaction potential; and to improve the soil to give adequate strength.

Conventional bearing capacity calculations are of little use with complex loading and geometry as they depend heavily upon geometric idealizations. Resort was made to finite element analyses using a program developed at Imperial College, London specifically for the analysis of plane-strain geotechnical problems. This allowed the non-linear behaviour of the material to be modelled and the construction sequence to be simulated. The behaviour of the breakwater under different assumed soil strength parameters allowed a minimum acceptable soil strength to be selected. Fig. 2 shows the design forces on the structure, together with the assumed soil profile after dynamic consolidation had taken place. The insets show local factors of safety (F_L) at design liveload and at 1.4 times liveload. For failure, a continuous zone must exist where $F_L < 1.0$ and it can be seen that even at the higher liveload this has not quite occurred. These diagrams relate to an assumed $\emptyset'$ value of 36° (with corresponding values of tangent modulus and Poissons ratio) which was the selected minimum value which had to be attained by the dynamic consolidation process. At this strength the most likely failure mode is a shallow sliding failure, whereas lower strengths showed the development of a deep-seated rotational failure. Following laboratory tests, a critical state line was postulated for the very silty sand (Fig. 3). Soil whose natural in situ condition is represented by a point significantly above the critical state condition will be particularly liable to liquefaction as a result of repetitive or pulsating stress, such as wave action, where the times available for dissipation of positive pore pressures are limited. The very silty sand is such a material and, in its untreated condition, positive pore pressures of about 35 kPa could be expected to be generated under wave action. These would certainly have resulted in the collapse of the structure. From the postulated critical state line it was anticipated that the moisture content of the very silty sand layer would have to be reduced by dynamic consolidation from its untreated value of about 45% to no more than 33%.

A dynamic consolidation trial was set up on an area 100 m by 40 m, which at the time (winter 1977) is understood to have been the first such operation in open sea conditions, although the technique had been proved in sheltered waters (Fig. 4). The 32 t weight, which was voided to minimize losses on impact with the sea, was dropped from a height of 10 m above the sea into depths of water varying up to 7 m, onto a rock blanket spread on the sea bed. A 4 m grid was used between craters and it was found that maximum settlement was achieved at no more than 10 blows. A second pass was then made of 7 blows on a 2.5 m grid giving a similar energy input. Induced settlements measured on a grid of points after the trial were of the order of 500 mm. Certain practical points from the trial are worthy of note. The first is that only the crudest assessment of induced settlement could be made during the trial, based upon marks on the drop wire. Secondly, piezometers were installed in an attempt to measure the rise in pore pressures and also the rate of dissipation.

Breakwaters—design and construction. Thomas Telford Ltd, London, 1984

79

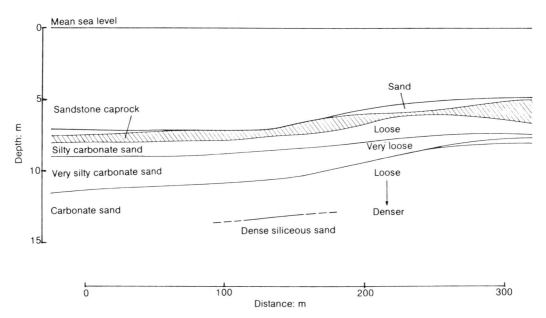

Fig. 1.   Offshore section of breakwater

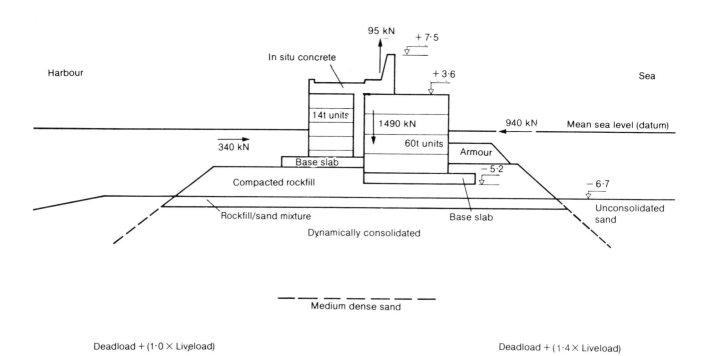

Fig. 2.   Breakwater section and loadings

In land-based work the piezometer leads can be protected and buried, and locations chosen which are between craters. In this over-water work, although armoured sheaths were used beneath the rockfill, less protection was possible and the drop-zone was less well controlled. Consequently, none of the piezometers survived the second pass, although a limited amount of data was derived from the first pass, particularly the rate of dissipation which indicated that equilibrium conditions had been achieved within one week of completion of the first pass. Finally, quite substantial periods of down-time occurred due to the weather conditions, as it was found that even quite moderate seas produced sway on the pontoon which prevented the safe dropping of the weight. It therefore seems unlikely that the process could be used reliably in offshore conditions which are more exposed than those encountered in the relatively sheltered waters of the Arabian Gulf.

The results of the trial were assessed using four types of test: the SPT, the Menard Pressuremeter, the in situ moisture content, and triaxial tests on samples recovered using the Bishop Sampler. Although the SPT showed encouraging results in the less silty sands above and below the very silty layer, the N value in this layer was still only 3 blows, which seemed disappointing. However, most SPT correlations have been made in clean silica sands and it has been suggested that saturated silty sands may give very low values of N even when in a medium-dense state and also that carbonate sands would, by virtue of lower skin friction values, give lower values of N than silica sands of the same angle of shearing resistance. It was, therefore, decided to treat the SPT results in the very silty sand layer with some caution.

The results of the pressuremeter and moisture content tests are shown in Fig. 5 and indicate a significant rise in pressure limit through both passes, but no apparent improvement in water content after the first pass. Although the change in pressure limit was encouraging, there did not appear to have been quite enough improvement in water content to guarantee stability. Therefore, in order to determine whether the postulated critical state line had been set too conservatively or whether the material would exhibit liquefaction phenomena under cyclic loads, it was necessary to examine closely the behaviour of the soil in the triaxial test. From the consolidation phase of the tests it was apparent that dynamic consolidation had caused the soil to become less compressible; the static tests showed that $\emptyset'$ was well in excess of the required value of $36°$ and that, at the mean stress level in the very silty layer, the behaviour of the soil was on the borderline between contractive and dilatant behaviour at peak shear stress. The cyclic loading tests showed that at the same mean stress level no signs of liquefaction were exhibited even after 120 cycles, while the average ratio of pore-water pressure to mean effective stress at the end of cycling was 0.45. Dissipation of this pressure will cause further small settlement. It was concluded on the basis of all the tests that dynamic consolidation had been successful in achieving the required foundation behaviour

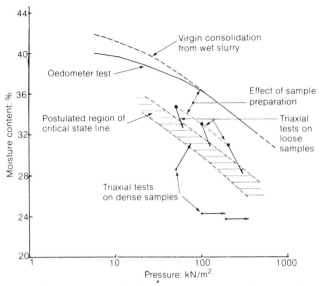

Fig. 3. Pressure/moisture content relationship

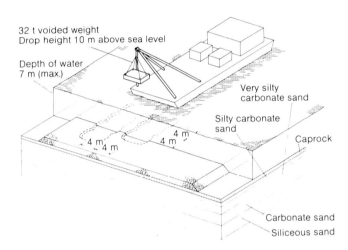

Fig. 4. Diagrammatic representation of dynamic consolidation of sea bed

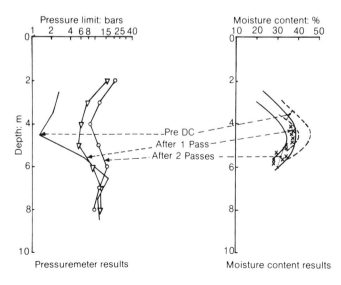

Fig. 5. Pressuremeter and moisture content results

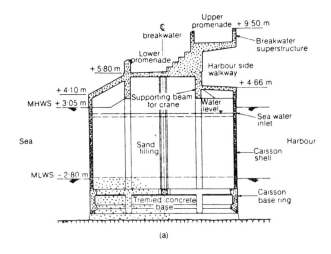

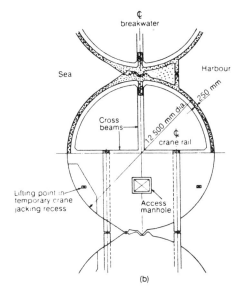

Fig. 6. Details of caisson-type breakwater at Brighton Marina

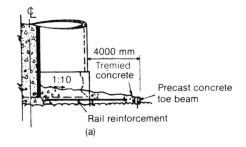

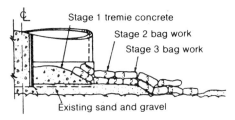

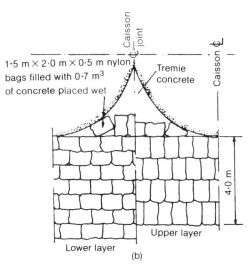

Fig. 7. Alternative methods of scour protection – diagrammatic view

at this site and that, although this was only just true of the very silty layer, its behaviour would improve under the action of tidal variations in weight and of wave action, at the possible expense of very minor settlements of the structure. It is, however, clear that the very silty sands were on the borderline of soils which can be improved by this process.

Dynamic consolidation was used on the whole of the breakwater foundation with control testing by pressuremeter and moisture content. The first part of the breakwater was built on the trial area and the blocks were surcharged to simulate the in situ core and part of the liveload. Settlements of no more than 15 mm were observed and these were complete within one week of loading. During construction the breakwater was hit by a storm of about half the design loading. Close inspection showed no signs of movement or distress and none has occurred subsequently.

Offshore dynamic consolidation is a feasible technique which may be valuable for difficult sites, provided that the sea state is moderate enough to permit the weight to be dropped safely and that the soils to be improved are

permeable enough to demonstrably safe end results.

P. GANLY (Lewis & Duvivier)
With the magnitude of the capital costs involved in construction and the serious consequences of subsequent failure, the Author is quite right to highlight the necessity of providing effective measures to protect the foundation of the breakwater against erosion to preserve stability. This is especially so in the case of gravity walls which, unlike rubble mounds with their greater flexibility, are less able to tolerate uneven foundation settlement without impairing the serviceability.

In paragraph 22 the Author prompts comment upon methods of dealing with the problem of erosion which are based upon experience rather than theory. My contribution to the discussion illustrates the method adopted on a circular caisson-type breakwater, founded on chalk on the south coast of England, with which my firm was involved for the Brighton Marine Company Ltd. (Fig. 6)

After the trials in the early days of construction using sand-filled nylon bags, rock

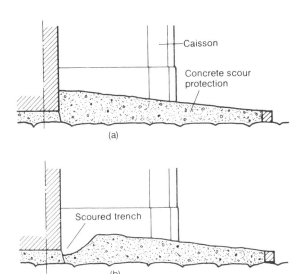

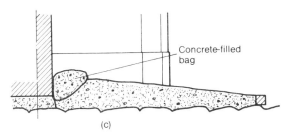

Fig. 8. Maintenance of scour protection – diagrammatic view

and asphalt and concrete with rail reinforcement, the latter method was chosen and for reasons of economy it was limited to what was considered to be the most vulnerable parts of the breakwaters.

The specification provided that the protective apron to each caisson be completed not more than three months after the caisson was placed, this timing to be varied at the Engineer's discretion depending on physical conditions experienced in practice.

It was fortunate that this proviso was included as during an exceptionally heavy storm while construction was in progress, and without the protection of the apron, the chalk foundation was scoured away to such an extent that three of the caissons settled by up to a maximum of 650 mm

In the aftermath of what could have been a catastrophe (the crane which was constructing the breakwater was marooned on the next two caissons), the design of the scour protection was immediately modified and a much more rapid method evolved using 1.5 m x 0.5 m nylon bags partially filled with 0.7 m³ of concrete placed wet by a crane standing on the breakwater (Fig. 7).

From then on we made sure that scour protection kept pace with construction and, accepting that substantial extra cost would result, a decision was taken to extend the protection over the total length of the breakwaters rather than the limited length initially contemplated.

Apart from minor maintenance no further problems were encountered during the remaining period of construction. Now, some 8 to 10 years later, with access for heavy lifting plant severely restricted, a rolling programme of

Fig. 9. Maintenance of scour protection - mattress and arrangement for filling

maintenance is in operation and after various trials a method using large fabricated mattresses placed by a diver and pumped full of concrete on the sea bed has been developed (Figs 8 and 9).

Progress, as has always been the case, is relatively slow being so heavily dependent upon the weather and conditions on the sea bed where the divers have to work.

The cost of the initial measure amounted to about 6% of the cost of the breakwaters at the time of construction.

The cost of maintaining the apron is working out at about £2700 per caisson. Out of the Company's total annual budget for breakwater maintenance, some £80,000 has been set aside for the work this year, which will cover 30 of the 110 caissons (12.5 m in diameter), i.e. approx. £200 per metre, or overall 1/3 percent of the estimated present day capital cost of the breakwaters.

D. J. VANDENBOSSCHE (*Haecon N.V., Gent*)
The overall geotechnical stability analysis of a rubble-mound breakwater affects not only the foundation layers of the structure but also the internal stability of the mound itself. This stability analysis involves careful consideration of wave loads, pore pressures, wave impact forces, seepage, material characteristics (e.g. angle of friction, cohesion).

Instrumental observation of piezometric pressures has been used successfully also by monitoring the pressure waves in sandy layers due to wave action.

By preparing specifications, clear monitoring programmes of both execution and performance after completion of foundation works are made possible. Performance criteria and control measures should be established.

The monitoring of performance in the case of soil replacement should be executed prior to the breakwater construction.

I would like to stress the importance of proper site investigations to conduct a liquefaction risk analysis when sandy bottoms are encountered.

H. ALTINK (*Delta Marine Consultants, The Netherlands*)
My company has evaluated various solutions to the problem of placing breakwater on very soft subsoil for a marina project in Venezuela. We compared massive soil improvement, partial improvement by large diameter sand piles and also a profile including·a tensile fabric with the 'standard' solution of flat slopes. We concluded that an appreciable increase in safety to sliding could be achieved by the methods mentioned and that for the specific case the massive soil improvement worked out to be most feasible after cost comparison. However, in other cases the methods stated may have to be renewed.

J. G. HAYMAN-JOYCE (*Livesey & Henderson*)
Some features of the construction of the 1 km long rock armoured rubble-mound breakwater completed in 1979 for the international fishing port at Vacamonte, Panama, may be of interest in relation to Paper 5.

The design and supervision of construction of this breakwater was undertaken by Livesey & Henderson who, in association with Pistsa-Greiner, a Panamanian consulting group, were the consultants for the whole project.

The design of the breakwater had to take account of the layer of very soft silt known as 'Pacific Muck' with depths from 5 to 13 m along the line of the breakwater, overlying decomposed and hard basalt rock. This soft silt had negligible shear strength, but the decomposed basalt formed an acceptable foundation strata. Two main alternative methods of carrying the breakwater loadings to the decomposed basalt were investigated: replacement of the soft silt under the breakwater with suitable sand fill, and displacement of the soft silt with rock fill. Analysis showed that the former method would be significantly less costly and more reliable.

Analysis of available sand from nearby Punta Chame showed it to be a medium coarse sand with a narrow range of particle sizes. The possibility of liquefaction of this sand under seismic tremors was extensively investigated. The potential for liquefaction of the sand was considered to be borderline at a seismic acceleration of 0.16 g. Analysis of the earthquake statistics for the Panama area showed that a seismic acceleration of 0.1 g is unlikely to be exceeded in 200 years and an acceleration of 0.16 g is unlikely to be exceeded in 2000 years.

The trench under the breakwater was dredged with an 8 m^3 grab dredger and temporary dredged slopes of 1:1 proved stable provided a sand blanket was placed within 7-10 days. In practice, the sand replacement filling followed closely behind the dredging; a typical tug/barge operational cycle of loading 1000 m^3 dredged spoil, dumping the spoil 15 km out to sea in deep water, loading 800 m^3 sand fill at Punta Chame (where another 8 m^3 grab dredger was stationed) and dumping in the dredged trench averaged 5 hours.

A blinding layer of quarry run rock up to 250 kg was placed over the sand fill to a depth of 1 m and the core was subsequently placed by end tipping. Toe protection was placed by an 800 m^3 capacity vessel using a 2 m^3 grab.

Breakwater construction progressed at an average of 10 m per day, including placing the core rock, underlayers and armour rock. A 60 t crawler crane placed the primary and secondary armour rock, and placed the capping rock and harbour side armour rock as it worked back from the breakwater.

Tests of the sand filling were carried out with a Dutch Cone Penetrometer apparatus, and based on estimates made of the probable settlement when loaded with the breakwater rock, the Contractor allowed for this settlement when placing the sand. As placing of the core material progressed, a number of settlement plates were positioned on top of the sand fill to allow monitoring of the sand and superimposed rubble-mound settlement as construction proceeded. The settlement plates, 1.2 m x 1.2 m x 0.012 m thick, had rods welded to them which extended to final crest level, within 0.15 m diameter casing acting as a loose sleeve and protective cages.

Measurements of plate levels were taken at regular intervals and showed settlements of

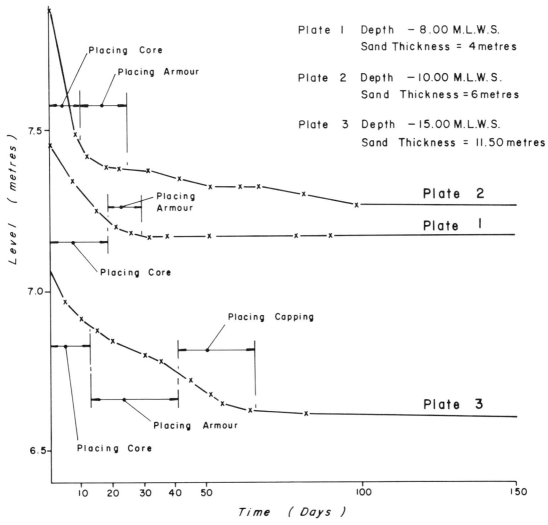

Plate 1   Depth   − 8.00 M.L.W.S.
Sand Thickness = 4 metres

Plate 2   Depth   −10.00 M.L.W.S.
Sand Thickness = 6 metres

Plate 3   Depth   −15.00 M.L.W.S.
Sand Thickness = 11.50 metres

Fig. 10.   Vacamonte settlement plates

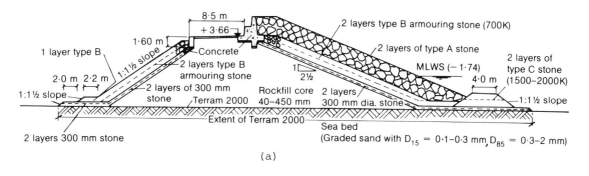

(a)

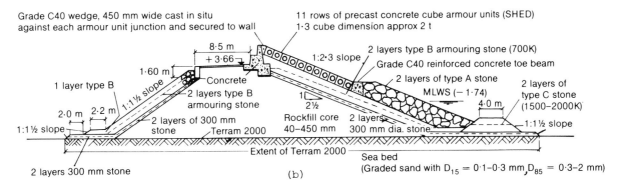

(b)

Fig. 11. Bangor − north breakwater: (a) original stone armouring section; (b) present SHED/stone armouring section

between 4 and 10 cm per m depth of sand fill placed. Between 25% and 75% of this settlement occurred during the placing of the core material, and the greater part of the remainder during and shortly after the placing of the underlayer and armour rock. About 5 months following completion of the breakwater head, settlement reduced to less than 0.05 cm per month and after approximately 8 months no further movement was recorded (see Fig. 10).

A further point of interest in the construction of the project was the successful use of a Terram filter to prevent sand filling placed in the reclamation areas leaching through the rockfill and rock armoured bunds which formed the seaward perimeter of the reclamation areas within the harbour.

I. M. GOODWILL (University of Leeds)
Armour units, when packed together, form a structure which contains a large number of voids of varying sizes. Some of the voids will be quite small and therefore, when subject to wave action, the flow of water through them will be at a high velocity. This is the condition necessary for cavitation to occur. Model tests suggest that the pressure of a wave following on the breakwater may be as high as 20 atm. If this is the case or even if the pressures are only half this volume, then cavitation will almost certainly occur in the smaller voids. This will cause severe pitting of the concrete and will eventually cause the concrete to fail when it is subjected to waves of relatively modest size. Such a condition would never be seen in a model test because cavitation will only occur at full-scale velocities.

What are the Authors' views on the suggestion that cavitation may be an additional cause of failure of armour units?

A. K. BELL (Kirk, McClure & Morton, Belfast)
My comments relate to the use of a composite breakwater armouring system using a combination of rock and regularly placed precast concrete units. The particular project for which we are using this form of armour is the new Harbour at Bangor, Northern Ireland.

The north breakwater is currently under construction with the second breakwater at the design stage. The breakwaters stand in 8-9 m of water at M.H.W.S. and were originally designed and model tested as an all rock armoured structure. A low rate of rock armour production necessitated the change, during the construction period, to the composite section of SHED blocks on the upper slope and rock armour for the lower slope (see Fig. 11).

I have carried out extensive model tests on the section at Queen's University of Belfast as well as a smaller study, using cob blocks, at Wallingford. The following points about the hollow cube type of block may be of interest.

(a) Arranging the rows of blocks to ensure that the vertical joints are staggered gives greater protection against rapid progressive collapse in the event of one or two blocks being damaged or removed. The staggered columns allow arching around the hole whereas straight columns allow the whole column of armour block to slide down the slope.

(b) Oblique wave attack produced significantly less overtopping than normal wave attack. It appears that the energy at the wave front can be dispersed along the breakwater rather than having to move the water up the slope. For example, at a wave angle of 25° to the normal, the overtopping volumes were only 10% of the volume measured with normal wave attack during tests with similar random wave trains.

(c) The concrete armour blocks must be well restrained at all the edges. This is of particular importance at round heads and other similar structures where the upper blocks can be easily ripped away.

Except in the case where there is little water depth at the structure at low tide, the greatest problem with this type of composite section is at the transition from rock armour to concrete unit. It is necessary for both aesthetics and the maintenance of interblock friction to have some form of toe beam to maintain a good line at the bottom of the slope of the concrete units. There is also the physical problem of changing from an armour layer thickness of say 1.3 m for the units to two layers of stone giving say a 3 m thickness.

Any beam at the toe must be stiff enough to spread the loads and bridge any voids created by movements of the individual armour boulders and yet be flexible enough to avoid cracking if the structure as a whole settles. During wave action a toe beam may also produce high local current velocities and back pressures which can dislodge armour boulders in the area of the toe beam.

Our solution to the problem has been to design and model test a voided toe beam to be cast at low tide level. The voided beam (approximately 50% void ratio) has been devised to attempt to maintain the permeability of the armour layer and to reduce the back pressures on the stone armour immediately to seaward of the beam. There is a central square void, from top to bottom, in the beam and a series of offset conical holes through to this from each face. The intention is to absorb some of the wave energy by turbulence within the beam. The beam is of reinforced cast in situ concrete ($40N/mm^2$) and has a joggle joint every 10.4 m.

We are currently investigating and testing alternative forms for this beam for the next breakwater. Included in these studies is the use of precast sections which have the advantage of good control of concrete quality but it is more difficult to ensure an even foundation loading and prevent leach out of material from beneath the beam.

There is no doubt that this problem of the concrete unit-armour stone interface can be solved by taking the concrete units down below the wave action zone. However, very practical problems then exist, with these regularly placed units, in achieving a sufficiently accurate grading of the underwater slope for bedding the units and producing a good line at the toe. It must be borne in mind that if the line of units is wrong at the toe, the error will be reproduced all the way up the slope. A poor line can lead to uneven packing of the individual units and the reduction of interblock friction.

I think it would sometimes be useful for designers to put on their wellington boots, get into a wave model basin, and actually try to build physical models of their proposals under the sort of conditions that are present on site.

T. SØRENSEN *(Danish Hydraulic Institute)*
The Authors of Paper 6 raise two points in their discussion of the modelling of the wave climate in laboratory tests:

(a) A length of wave records of 20 min is too small.

(b) Groupiness of the design storm might be different from that of the worst storm recorded.

Professor Battjes in Paper 1 claimed that the groupiness was fully defined by the wave spectrum. The insufficiency of the spectrum to describe natural waves is illustrated by Fig. 12. Curve (a) shows a natural wave train with high crests and shallow troughs, with steeper fronts and more gentle back sides. Curve (b) shows a most unnatural 'wave train' with low crests and deep troughs, with gentle fronts and steep back sides. Curve (b) is the mirror of curve (a) and they have the same spectrum.

If a wave train is reproduced mathematically from the same spectrum with a large number of sinusoidal components, you get at the wave generator a nice, round-crested product without vertical or horizontal skewness, as illustrated by curve (b).

Thirteen years ago, in connection with the model tests with the Brighton Marina Breakwater, the Danish Hydraulic Institute realized that such a mathematical wave train was physically unsatisfactory because it did not produce a single shock force on a vertical face breakwater. Since then, natural wave trains have been produced, whenever possible.

The differences between the two wave trains shown lie in the second order components. These components are bound to the higher waves. The proper wave shape will develop in a long flume with wind, but it requires quite a long flume, which costs money.

The present philosophy at the Danish Hydraulic Institute regarding wave reproduction in the laboratory is given in the May 1983 issue of Danish Hydraulics in the article 'Waves - nature and scale model'.

If you reproduce a recorded wave train that has been so transformed that its spectrum is identical to that of the forecast design storm you are all right according to the authors of Papers 1 and 6. But in addition, you should make sure that you reproduce a realistic groupiness and wave shape. Why run the risk of missing out on something that might be important but not yet fully understood?

J. D. METTAM *(Bertlin & Partners)*
In his introduction to Paper 6 the Author defined his main purpose as being to reproduce waves and structures as closely as possible. While this is an excellent aim for hydraulic research, I as an engineer aim to provide my client with a structure that is satisfactory in service.

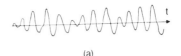

(a)

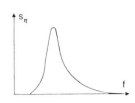

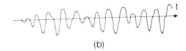

(b)

Fig. 12. Wave train

Mettam and Berry (ref. 1) concluded that the large number of breakwater failures results from a simple philosophical error in the use of model tests for design of breakwaters, and that the remedy is a simple change in modelling technique; that is, deliberately modelling armour units and cappings with a specific gravity less than in the prototype.

Our present philosophy has its origins in the history of breakwater design. Some old breakwaters, such as Alderney and Portland, with rock armour have been kept in service by continuous maintenance, new armour being tipped onto the breakwater to replace that broken or worn down or washed away. With a quarry close by and plenty of cheap labour this can be an acceptable solution still.

But with concrete armour, or large rock requiring special cranes, a more permanent solution is needed.

Because of development of our designs using hydraulic models, we have the position that the final tests of a design show that it suffers, for example, 1% damage under a design storm of 1 in 50 (or 100) years. If we build the design exactly as tested there will be a 63% chance of the design wave being exceeded during the 50 (or 100) year return period, and it could come next year. We are therefore constructing a design with a 63% chance of significant damage during 50 (or 100) years. We are deliberately building a structure on the verge of failure - even assuming our knowledge of waves and our techniques of modelling are perfect.

But our knowledge of waves and our modelling techniques are far from perfect and contractors do not normally build breakwaters exactly as designed even when the work is properly supervised by the designer (who may not in fact be employed on the site at all).

In most other designs we include in our design calculation a factor of safety to allow for human ignorance and error, but we do not normally do this when using model test results to check and finalize our designs for breakwaters. It would however be possible to do this by deliberately reducing the specific gravity of armour and capping in the model. If our flume tests show the predetermined degree of movement or damage under the design wave, we would know that we can build the prototype as modelled and that it will have a factor of safety against this degree of damage.

This would provide a more systematic margin of safety than can be obtained by making adjustments to the tested design to ensure a stronger prototype.

Application of this technique would also enable a more valid comparison to be made with other forms of construction, such as some forms of vertical construction, which are designed to normal structural codes and automatically include factors of safety (or in the case of limit state design partial factors of safety).

The result of adapting this technique would certainly be a reduction in breakwater failures, but it could also result in a different choice of type of structure.

D. J. VANDENBOSSCHE (*Haecon N.V., Gent*)
The Authors state that 'the primary objective of model testing....is to check the stability of the breakwater up to and exceeding the design sea state. However, modelling is also used to gather information on the hydraulic performance of the breakwater, in terms of reflection, run-up, overtopping and wave transmission'.

I should quote this procedure as the investigation of the serviceability state of the structure. I believe that hydraulic physical model tests should also include:

(a) Investigations to the limit-state conditions of the structure. By this, the mode of failure or severe damage could be investigated and observed thoroughly. These test series will give ample data to the risk evaluation.

(b) Hydraulic models with proper scales can also give calibration data on pore pressure, wave impact forces, etc. to numerical model analysis.

P. LACEY (*Ove Arup & Partners*)
I am interested in the concept described in Paper 6 of saying to contractors 'we may be designing a breakwater; shall we discuss it and then we can test it?' Is this practical? Should not the engineers be saying to clients 'we will not be testing your structure in the shortest possible time', but should be educating the clients into a testing programme which can assess the conditions and duration of wave attack, and more importantly, the likelihood of damage. After all, the early engineers took years in prototype to get breakwaters correct. Why now the indecent haste?

H. F. BURCHARTH (*Aalborg University, Denmark*)
A question was raised whether the short wave spectrum, together with random phases of component waves, uniquely specifies the wave grouping. We must distinguish between a mathematical description and a physical representation.

For a mathematical (theoretical) description we need more than the short wave spectrum and random phases since an upside-down turning of the amplitude time series does not change the short wave spectra. Nor is the second order long wave spectra changed by such an operation. Therefore, some phase restrictions must be put on at least the higher-frequency range component to assure wave asymmetry. In physical models, however, generation in accordance with the short

wave spectrum and random phases seems sufficient (at least for a first approximation of wave group reproduction) because higher order harmonics, which create the asymmetry, are generated automatically.

W. R. THORPE (*Paper 5*)
Throughout Mr Quinlan's account of the design and construction of the Kuwait breakwater foundation the importance of soils and wave data is apparent. The correlation between design and the strength of the consolidated loose silty sand rested with sampling and testing interpreted with skill. It is also clear that wider application of dynamic consolidation as an offshore construction method will not occur unless it can operate in the same sea conditions as other floating plant.

Similar dependence on sea states and on reliable forecasts of waves are evident in Mr Ganly's case history of the measures to combat scour at the Brighton Marina breakwater. Details of scour protection methods and their costs are seldom released and all maritime engineers will be grateful to Mr Ganly for the facts he has published.

With reference to Mr Vandenbossche's contribution, pore pressures and wave impact forces and their interrelationship with material characteristics have not been investigated thoroughly despite general recognition of their importance in the overall analysis of stability. If, as a result of suggestions put forward at this conference, the necessary model studies begin, the results may show that the operational wave is no less important than the extreme event.

Mr Hayman-Joyce's description of the Vacamonte breakwater foundations shows that, given the freedom to dump spoil at sea, deep soil replacement can be carried out economically.

Coincidentally massive soil replacement emerged as the economical solution to the breakwater founded on very soft subsoil described by Mr Altink. However, as Mr Altink states, each foundation problem must be judged on its own merits and also on a judgement of the equipment and materials available, as well as the impact on the environment.

M. W. OWEN and **N. W. H. ALLSOP** (*Paper 6*)
The Authors agree with Dr Goodwill that it seems more than likely that cavitation occurs within some of the voids formed when a large number of armour units are packed together to form a structure. Cavitation must therefore play some part in the failure of armour units. However, the Authors have not observed any evidence of cavitation damage on full-sized armour units. It seem likely that cavitation is a relatively minor cause of failure, compared to impact loading on the armour units for example.

The Authors thank Mr Bell for his very interesting contribution describing the use of hollow-block armour units at Bangor, N. Ireland. They find themselves in complete agreement with the various practical points given for the use of this type of armour unit, and will watch with great interest the performance of this particular breakwater over the coming years.

Referring to Mr Bell's final comment, and also to the comments made by Mr Lacey, the Authors agree wholeheartedly that all those involved in a breakwater project should work much more closely to achieve the optimum design. This means that we should attempt to educate the client to accept much more reasonable time scales for feasibility and design stages. We should encourage the designer to liaise more closely with the laboratory, especially in the early stages when the basic philosophy of the breakwater is being decided and preliminary ideas are being tossed around. Also we should involve the contractor as soon as possible so that we all have a clear idea of what can and cannot be built at full scale.

Mr Mettam is concerned with the application of model results in breakwater design, and recommends a deliberate modification to the model breakwater. By designing the model breakwater to be stable a factor of safety is thereby introduced for full-size breakwaters. He advocates that this modification be achieved by reducing the specific gravity of the armour units and capping. Of all the possible parameters which could be modified, this particular parameter is presumably chosen on the grounds that it achieves a reduction in the weight of the armour units without in any way distorting the breakwater geometry, and hence its hydraulic performance (overtopping, reflections etc.). Unfortunately, it is by no means clear to what extent the stability of the breakwater is distorted by reduced specific gravity. Although the various formulae for determining the weight of the required armour unit all include specific gravity as a parameter, to the best of the Authors' knowledge tests to determine the precise effects of differing specific gravities have been carried out only for rock armour and for dolosse (ref.2). For both these armour types it seems probable that the effect of changing specific gravity is that less than the various formulae would predict. For the various different armour types it is thus not possible to say what safety margin is introduced by artificially reducing the specific gravity in the model armour units.

In the opinion of the Authors, a better approach is to test the breakwater at wave heights significantly greater than the design wave condition and if possible testing to destruction, noting the mode of failure or of severe damage. This will then give a good idea of the safety margin available in the break-

water design, and will enable the consequences of failure to be assessed, both in terms of performance of the remaining cross-section, and of possible reconstruction methods. This approach would tie in closely with that suggested by Mr Vandenbossche in his contribution. The Authors would also agree with him that properly designed hydraulic models of breakwaters can be used to provide calibration data for numerical models.

Mr Sørensen questioned the sufficiency of the wave spectrum to fully describe natural waves, in terms of both wave shape and wave groupiness. He showed two wave trains having the same spectrum, but with markedly different shapes; one wave train being an upside-down image of the other. To a large extent Mr Sørensen's question has been answered in Professor Burcharth's contribution. In hydraulic models it is impossible to generate upside-down waves: the basic physics of surface waves does not allow it, thus demonstrating one of the advantages of physical models over numerical models, where almost anything is possible, however unreal. Turning to wave groups, neither the Authors nor Professor Battjes in Paper 1 were claiming that the wave spectrum alone was sufficient to describe wave groupiness. The available evidence suggests, however, that all that is necessary to correctly reproduce the probability of occurrence of wave groups is to allow random phases in the generation of the required spectrum. In models we must then generate sufficiently long wave trains for those wave groups to show themselves. If a natural recorded wave train is used in model tests the wave phases are exactly prescribed, so the wave groups occurring may not be fully representative of the true sea state, unless exceptionally long records are available. In some cases it is therefore possible that the use of recorded wave trains in model tests may underestimate the occurrence of the worst combinations of waves causing damage to the breakwater.

REFERENCES
1. METTAM, J. D. and BERRY, J.G. Factors of safety for the design of breakwaters. 18th ICCE, Capetown, 1982.
2. SCHOLTZ, D. J. P., ZWAMBORN, J. A. and VAN NIEKIRK, M. Dolos stability-effect of block density and waist thickness. Proc.18th Int.Conf. Coastal Engng, Cape Town, November 1982.

7 Field scale studies of riprap

P. ACKERS and J. D. PITT, Binnie and Partners

SYNOPSIS
The performance of five panels of riprap was observed at an offshore, tidal site in the Wash for
$2\frac{1}{2}$ years after their construction, by which time all but one had failed. Comparisons were made of
the damage histories of the riprap panels with those estimated using laboratory data. These did
not confirm a scale effect, causing riprap sized on laboratory data to be larger than necessary.
Observed damage was marginally greater than would have been predicted by scaling from laboratory
tests. There were differences, however, between the laboratory conditions and the coastal situation in the Wash, and in the characteristics of the slope protection, that may have contributed to
increased vulnerability. The sensitivity of stability not only to wave height and stone size but
also to construction methods is apparent.

INTRODUCTION

Use of riprap for slope protection
1. Graded quarry stone is frequently used to
protect the surface of earth embankments from
wind-generated waves. It is an alternative to
concrete slabs or blockwork, bituminous surfac-
ing and randomly dumped or pattern-placed
concrete units, and the frequency of its use
demonstrates that in many situations it is more
economical material than the alternatives. A
common situation for the use of riprap is on the
upstream face of an embankment dam, where the
waves will approach over deep water, usually
with a relatively short fetch.

2. Coastal situations differ in that they are
tidal, tend to be shallower and are exposed to
longer fetches. Coastal embankments where riprap slope protection is used on an impermeable
foundation should be distinguished from rubble
mound breakwaters, which are characterised by
their permeability and sometimes by overtopping,
as they are not water-excluding or retaining
structures. The use of stone for breakwaters is
governed by similar criteria, but the work reported here relates specifically to coastal embankments with an impermeable core.

3. Most design methods for riprap are based on
the results of model tests carried out in a
hydraulics laboratory. The validity of these
methods depends on how well the models reproduce
all the characteristics of the full-scale riprap
and its filter layers, as well as the wave climate. It also depends on whether there are
'scale effects' arising from hydrodynamic aspects of the performance that can not be scaled
down, and the validity of any adjustment consequently made in extrapolating from laboratory
to field conditions.

Previous research
4. In 1962, the Civil Engineering Research
Association (as the Construction Industry Research and Information Association, CIRIA, was
then known) sponsored laboratory tests at the
Hydraulics Research Station, Wallingford (HRS),
which resulted in the publication of a Report
(ref. 1) which was widely used for the design
of slope protection. In that work, results of
the main series of tests using regular waves
were related to some tests in a wind-wave flume
with a natural spectrum of waves. Work on the
subject was continued at HRS in collaboration
with CIRIA, using paddle-generated irregular
waves, culminating in the comprehensive CIRIA
Report 61 (ref. 2), which updated the UK methodology.

Objectives of field study
5. The object of the study reported here was
to observe at full scale the actual performance
of riprap slope protection under attack from a
natural wave climate, in order to compare
damage with what would be deduced from laboratory tests at much lower Reynolds number. An
opportunity arose with the construction of an
offshore embankment within the Wash estuary as
part of the studies of water storage in that
area. The site was subject to variable water
levels, drying out at low water of most tides.

FIELD STUDIES

Design of experiments
6. The information available when designing
the experiments in 1974 was:

- about a year's measurement of wave
  heights at three sites in the Wash (see
  Figure 1 for locations)

- the results of HRS tests with irregular
  waves in the form of damage charts
  (ref. 3)

- the Coastal Engineering Research
  Centre's (CERC) conclusions on scale
  effects (ref. 4).

Breakwaters—design and construction. Thomas Telford Ltd, London, 1984

91

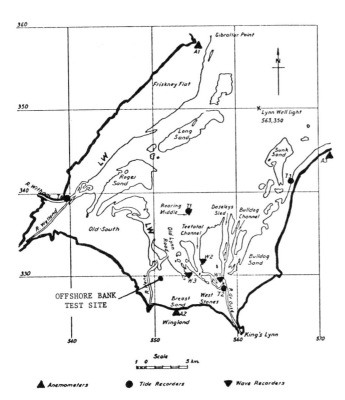

FIG. 1.   Location of experimental site and data acquisition stations.

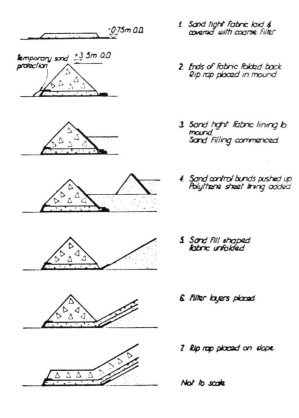

1. *Sand tight fabric laid & covered with coarse filter.*

2. *Ends of fabric folded back. Rip rap placed in mound.*

3. *Sand tight fabric lining to mound. Sand filling commenced.*

4. *Sand control bunds pushed up. Polythene sheet lining added.*

5. *Sand fill shaped. Fabric unfolded.*

6. *Filter layers placed.*

7. *Rip rap placed on slope.*

Not to scale.

FIG. 2.   Construction sequence of offshore bank.

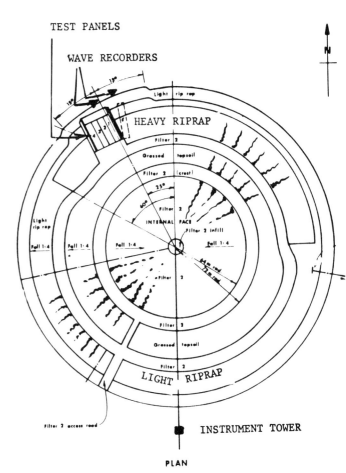

FIG. 3.   Location of test panels on offshore bank.

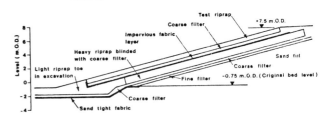

FIG. 4.   Section through test panel

TABLE 1.   Dimensions of riprap and filter layers.

	Panel 1	Panel 2	Panel 3	Panel 4	Panel 5
RIPRAP					
Specified D_{50} (mm)	200-250	300-350	400-450	550-600	650-850
Measured D_{50} (mm)	230	400	500	560	660
Layer thickness, t (mm)	440	480	570	760	1320*
t/D_{50}	1.92	1.21	1.14	1.35	2.0*
FILTER LAYER					
D_{50} (mm)	40	40	40	40	40
Layer thickness (mm)	380	380	390	430	300*+

* design values

+ also 200-mm layer of fine filter underneath.

7. For positive results, damage to all the sizes chosen for study (and perhaps failure of one of them in a typical year) was desirable. The uncertainty over scale effects posed a problem in selecting a range of sizes: those chosen had to give useful results whether scale effects in the HRS laboratory tests were nil, were as large as CERC had concluded or were somewhere between. The waves to be expected during a year's experiments were also unknown.

8. Damage calculations were repeated with and without allowance for scale effects for stone sizes of 100 mm, 200 mm, 300 mm, 460 mm, 650 mm and 850 mm and the results were considered on the basis of:

- a year's damage excluding the most severe observed wave event, with maximum scale effect assumed

- the same but with no scale effect

- the damage in the most severe observed event (a possible but unlikely first storm)

- the damage in the next highest wave event (a likely first storm).

On this basis, the following stone sizes were considered desirable for four test panels:

1. $200 < D_{50} < 250$ mm
2. $300 < D_{50} < 350$ mm
3. $400 < D_{50} < 450$ mm
4. $550 < D_{50} < 600$ mm

9. Four riprap test panels were constructed, each 6 m wide and approximately 26.5 m long, on top of the heavy riprap (HRR) that forms the basic surface protection to the bank. The panel width had to be limited for reasons of economy. Proposals for strengthening the edges of the panels, so that if one panel failed completely the adjacent panel would not be weakened, were considered. However, any method employed would then form an upstanding edge, which could cause undesirable wave reflections or could itself be washed away. Bearing in mind the cost and uncertain performance of any such arrangmeent, no special edge treatment was incorporated.

Construction
10. The offshore trial embankment on which the riprap test panels were placed was built on the Wash foreshore in 1975 at a site just east of the Wisbech Channel about 4 km from the sea-defence bank (Fig 1). It is ring shaped, but encloses only a small area of foreshore. It consists primarily of sand fill dredged from a nearby borrow pit. The foreshore level at the site is about −0.75 m OD, and the bank was built up from this level with side slopes of 1 in 4 to an 8 m wide crest at +14.0 m OD. The external diameter of the bank at foreshore level is approximately 280 m. The outside surface of the bank is protected against wave attack by riprap placed on two filter layers. Two sizes of riprap were used for the slope protection

of the main embankment. The smaller size, designated Light Riprap (LRR) was placed on the southern half of the bank to a top level of +6.0 m OD while the northern more exposed half was protected with Heavy Riprap (HRR) to a top level of +7.5 m OD. In order to prevent erosion at the foot of the slope a 10 m wide toe was constructed of LRR placed on a layer of coarse filter overlying a layer of sand tight fabric. The construction method is illustrated in Fig.2. It was this embankment that provided the foundation for the test panels.

11. The permanent HRR was blinded with filter material to ensure that no settlement of the test panels would occur through the filling of voids beneath them. A fabric sandwich, consisting of two layers of a non-woven and sand-tight fabric separated by a sheet of PVC, was placed on the blinded surface to make the test panel foundation impervious, matching the previous laboratory arrangement, and a layer of filter stone was placed over the fabric. The stone forming the test panels was placed on this layer, with the panel containing the smallest stone (Panel 1) at the eastern edge of the test area, each panel being flanked on its western edge by the panel containing the next larger size of stone. The location of the panels is shown in Figure 3, and a longitudinal section through one of the panels is shown in Figure 4.

12. In addition to the four panels for the CIRIA tests, an area (designated Panel 5) of the main bank protection immediately to the east was selected for monitoring. Construction of the test panels revealed several problems of control during placing, including the tendency for larger sizes to seggregate out, leaving uneven distribution of rock size. Although some adjustment was possible, the final gradings, the median stone sizes for the panels and **the layer thicknesses** were found to differ somewhat from the intended values (Table 1).

Instrumentation
13. Two wave pressure transducers were mounted on short piles, just beyond the edge of the submerged toe in front of the test panels, the timing and recording units of these being located in an instrument cabin supported on piles to the south of the embankment, well above maximum tide and wave level. This instrumentation was supplied, installed and serviced when necessary by the Institute of Oceanographic Sciences. The timer switched the recorders on alternately, each operating twice for 16 minutes around the high water period, with gaps of 60 minutes between the four recording sessions. This system was devised to give the maximum coverage of the wave climate over a high-water period consistent with keeping the interval between visits to change wave recorder charts to about 1 week.

14. Water level and other environmental observations were made, at locations illustrated in Fig. 1.

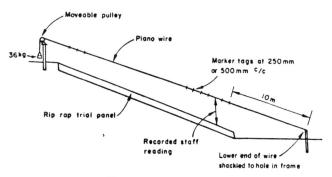

METHOD OF SETTING UP WIRE

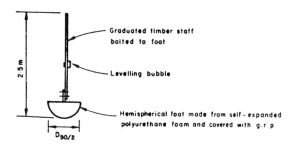

DETAIL OF MEASURING STAFF

FIG. 5.   Survey method

TABLE 2.   Comparison of measured and predicted damage in selected events (see ref. 5 for full details).

Event			Panel 1	Panel 2	Panel 3	Panel 4	Panel 5
1, Nov. 1975	$H_s = 1.0$ m $\bar{T} = 4.4$ s	Measured damage	839	83	\multicolumn{3}{c}{Not measured}		
		Predicted damage	313	32			
33, Aug. 1977	$H_s = 1.0$ m $\bar{T} = 4.1$ s	Measured damage	F	231	18	35	27
		Predicted damage	F	68	19	9	2
42, Jan. 1978	$H_s = 2.1$ m $\bar{T} = 6.1$ s	Measured damage	F	F	236(F)	188(F)	27
		Predicted damage	F	F	146	68	32

Notes:
1. Damage measured as equivalent number of D_{50} size stones eroded
2. The maximum error associated with measurement of damage is estimated at N = 39, 26, 21, 17, 12 for Panels 1, 2, 3, 4 and 5, respectively
3. F indicates failure.

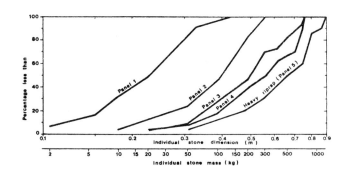

FIG. 6.   Grading curves for rip-rap.

TABLE 3. Comparison of calculated and field assessment of stone size (cm) for stability and failure.

Wave event no. 1. 16-18 Nov. 75 $H_3 = 1.03$ m		Wave event no.42 11-12 Jan. 78 $H_3 = 2.10$ m		Notes and methods
Stable	Fail	Stable	Fail	
36 (47)	30 (40)	73 (97)	61 (81)	(a) CERA ref.1
41	24	73	49	CIRIA ref.2
35	23	72	47	SPM ref.7
30	22	61	45	CERC ref.4
34	28	71	57	CERC ref.8
45 (b)	35 (c)	66 (d)	58 (e)	FIELD TESTS

Notes.

(a) Based on layer thickness of 2 D_{50}. Bracketted figures are for layer thickness 1.5 D_{50}.

(b) Panel 2, D_{50} = 40 cm, severe damage; panel 3, D_{50} = 50 cm, very stable.

(c) Panel 1, D_{50} = 23 cm, total failure; panel 2, D_{50} = 40 cm, severe damage.

(d) Panel 4, D_{50} = 56 cm, failed; panel 5, D_{50} = 66 cm just stable.

(e) Panel 4, D_{50} = 56 cm, failed; panel 5, D_{50} = 66 cm, just stable.

Riprap surveys

15. To avoid discrepancies arising from differences between the methods of surveying the riprap in the laboratory and in the field, the survey technique chosen was similar to that used by HRS in the CIRIA study(2). The surface of riprap is very irregular and special methods were needed to define the surface well enough for successive surveys to identify and quantify movement. An essential feature is that very many spot elevations were taken at chosen and repeatable locations on plan. The surveying system used a probe with a hemispherical foot of diameter $D_{50}/2$.

16. Five survey lines, 0.75m apart, were set up on each panel, and stone levels measured at fixed points along each line. On Panels 1 and 2, these fixed points were 0.25 m apart while on the two larger panels and the HRR panel the intervals were 0.50 m. The survey lines were fixed by stretching a 17 gauge piano wire, tagged at the required intervals, from a hole drilled in a frame welded to pipes jetted into the seabed at the toe of the panels to a pulley fixed to a second frame at the top of the panels. The technique is illustrated in Fig. 5.

Riprap and filter grading

17. The specification called for riprap having a median diameter within the stated band, no stones exceeding a particular size, shape being such that the longest dimension was not more than three times the shortest dimension, and the small end of the grading defined by a minimum figure for the lower 15 percentile (D_{15}). The gradings specified were not particularly narrow, but were believed to be reasonably in accordance with what was available economically from the quarry. It will be appreciated that the checking of the grading of riprap is not easy. Sampling can not be carried out ·by the normal procedures applied to fine grained material (e.g. successive quartering). Sieving is out of the question, yet, in most laboratory researches at small scale, materials have been defined by sieve size. Samples have to be treated as individual stones, most of which are so heavy that mechanical handling is needed. This not only poses problems on site, it also means that it is impracticable to expect a quarry to select and deliver riprap complying fully with a close specification.

18. After delivery of the material to the offshore bank site, a representative sample (about 15%) of the three smaller sizes was taken. Samples were also taken from the LRR (also used in Panel 4) and the HRR, although these samples were a much smaller percentage of the total volume of stone delivered in these sizes. The material in each sample was graded by weighing on a spring balance while for every fifth stone in each sample three orthogonal dimensions were measured. The mass grading curves obtained were converted to dimensional grading curves using the relationship suggested in Reference 4:

$$M = 0.65 \, \rho \, D_S^3$$

which is based on conversion of sieve gradings, where D_S is sieve size, to individual stone mass. The resultant grading curves are shown in Fig.6.

19. The shape characteristics of the stones forming each sample were expressed in terms of ratios of the upper, intermediate and lower dimensions and are plotted in scatter diagrams, to be found in ref. 5. These showed that the ratios of upper and lower dimensions exceeded 3.0 for only a small number of stones in each sample, but that this ratio tended to increase as the nominal dimension of the sample decreased.

20. The coarse filter material used beneath the test panels was the same as the filter 2 used beneath the main surface protection to the bank. The process of transporting the material from the stockpile and placing it in position inevitably led to some segregation Accordingly, samples were taken from the as-placed material beneath the main bank protection from areas selected from apparently coarser and finer than average sections. The relationships between the filter and the various trial riprap sizes were compared with the recommendations in the CIRIA Report(2). This showed that the material was not ideally suited to the riprap sizes at least by these recommendations, although other researchers are less rigorous and would permit the observed stone/filter size ratios.

Layer thickness

21. Layer thicknesses were deduced from surveys of the filter surface carried out before the test riprap was placed and of the completed riprap using the special tag-line surveys as already described. The mean thickness of the panels could not be determined accurately until after the first tag-line survey had been completed, by which time it was too late to remedy any short-fall. In fact, visual estimates of layer thickness proved to have been unreliable, and mean layer thicknesses expressed as a ratio of the D_{50} riprap size were less than expected. This was partly due to a shortfall of material available, partly due to penetration of the riprap into the filter layer, but also because the actual D_{50} values were in most cases above the average specified value. Layer thicknesses and other details of the panels as placed are included in Table 1.

Analysis of waves

22. The output from the two wave recorders at the offshore bank was analysed to find the significant height, $\overline{H}_{x3}$, and recorded mean zero crossing period, T_x (i.e. the height and period of the waves as recorded at the sensor head sited a depth x below the still water surface). Then these quantities were corrected for depth attenuation effects to give $\overline{H}_3$ and T. The calculation of $\overline{H}_{x3}$ and T_x was carried out by a method first suggested by Tucker (ref. 6) and modified by Draper (ref. 7). The correction for depth attenuation was calculated by a method proposed by HRS which allows for the variability of wave height and period, assuming that wave

spectra in the Wash are characterised by the Joint North Sea Wave Programme (JONSWAP) spectrum (ref. 8).

23. The waves recorded during the study were grouped into 45 wave events, defined as a period of time during which at least one record on each tide gave waves with $\overline{H}_3 > 0.5$ m. The majority of records have wave periods between 3 and 5 s, events with longer periods usually being associated with water levels above 3.5 m. The highest levels tend to be associated with winds from a direction which also provides a long fetch.

Assessment of damage

24. The data obtained from the riprap surveys were analysed by computer to give a quantitative description of the erosion damage sustained by each panel and surface profile plots of each line surveyed. The volumes of material eroded from each panel since the beginning of the study, and also since the previous survey, were obtained by differencing. The definition of erosion damage to a panel was the same as that used in the HRS laboratory studies (ref. 2): the mass-equivalent number of D_{50}-size spherical stones eroded from a $9D_{50}$ width of panel, considering only negative movements of the surface (i.e. lowering of the surface).

RESULTS AND ANALYSIS

Wave events and damage history

25. It was unfortunate that large waves (just over 1 m in height) occurred very early in the test programme during a gale lasting from 16 to 18 November 1975. The main effect was the total failure of Panel 1. The riprap and the underlying filter were completely washed away from the central section of the panel. The upper limit of damage, about 6m from the top of the panel, was marked by a near vertical face showing the riprap and filter layers. The failure of Panel 1 had been expected to occur during the course of the first winter, but its occurrence so early meant that no results on progressive damage were available for this panel. Considerable damage was also recorded on Panel 2 during the gale.

26. Erosion damage on Panel 2 was serious and was concentrated on the side closest to Panel 1. A considerable part was undoubtedly because of the loss of edge support. So that as much useful data as possible could be obtained from Panel 2, a sixth survey line was installed between line 5 and the edge of Panels 2 and 3, and the panel damage was analysed in two parts.

27. During December 1975 three wave events were recorded, but water levels were in general quite low and so there was little additional damage to the panels. The next severe gale occurred on 3 January 1976, when winds of up to 40 m/s (90 mile/h) were recorded. Waves on this occasion were not very large, partly because the wind did not reach its peak until some time after high water and partly because it was blowing from the west. Tide levels were high, however,

which resulted in further erosion at the top of **Panel** 1, causing the area of Panel 2 affected by loss of edge support to extend up the slope.

28. Three events with waves of about 0.75 m occurred between mid-January and mid-February 1976. Only the first event was with winds direct in line with the test panels, and damage to Panels 3 and 4 as well as further damage to Panel 2 was noted. Three further events were recorded before the panel surveys of May 1976. Only one of these events occurred with waves orthogonal to the panels, and on this occasion water levels were not very high so little damage was detected.

29. In the period from May 1976 to April 1977, the maximum significant wave height recorded was 0.83 m, but on this occasion the wind was from the northeast (70° from orthogonal). Two further events with waves of about 0.8 m were hindcast from wind data. One of these was with northerly winds and some minor damage was recorded. However, the amounts are comparable to the probable maximum error in measuring damage, so the results were not significant.

30. Data collection in the Spring of 1977 was marred by the loss of wave records from 19 March to 8 May, a period when northerly and northeasterly winds were dominant. However, hindcasting of wave events demonstrated that particularly severe wave attack ($H_3 = 1.4$ m) occurred immediately after the panel surveys of 5/6 April, and that waves were directly orthogonal to the bank: damage to Panels 2, 3, 4 and 5 was as expected.

31. Immediately following the August 1977 survey there was further severe wave action (4 days with H_3 up to 1 m at high tide): winds primarily from the north east, though they backed to north north west and resulted in the total failure of Panel 2 and further damage to Panels 3, 4 and 5. One further event, with waves nearly 1.0 m high, was recorded in December, but the wind was from the west and no damage was observed as a result of this event.

32. Waves recorded in the event of 11/12 January 1978, when winds were from the north and tide levels rose even higher than the catastrophic storm surge of 1953, were far in excess of the previous maximum ($\overline{H}_3 > 2$ m at one HW) and resulted in the total failure of Panels 3 and 4. Panel 5, the permanent slope protection, suffered very little damage if any.

33. The method used to hindcast the damage sustained by the test panels was a variation of that set out in ref. 3, making the appropriate allowance for the different water densities in the laboratory and fieldwork. The data in that Report were used to prepare curves relating damage to number of waves incident on the panel for differing values of the ratio H_3/D_{50}. These curves were extrapolated to 20 000 waves by assuming that the slope of the curves at 5000 waves (the limit of the laboratory research) remains constant, and beyond

20 000 waves no further increase in damage was assumed. The total damage sustained by a panel was obtained by summing the damage arising from waves in each 0.1 interval of wave height to 1.10 m; then 1.4 m, 1.7 m and 2.1 m to cover the extreme events. This treats the waves in each height interval in each event as if they were idependent occurrences, not influenced by the previous history of damage.

Conditions of study
34. The wave climate at the offshore bank during the period of study was far from ideal in terms of studying progressive damage. The first event, the largest recorded in the first phase of the research, came within 2 weeks of placing the slope-protection before it had been practicable for initial surveys to be carried out for the three panels with largest stones. Then the final event with waves of 2.1 m was so much larger than the previous maximum that Panels 3 and 4 failed together. The 2½-year period of observation contained some unusually severe events, and the January 1978 storm gave the highest water levels ever recorded. In respect of tide level and wave height this was close to the design condition for the permanent slope protection of the offshore bank. Although a different distribution in time of the severe events might have provided more positive results, the research was on the whole favoured by the severity of attack in a reasonably short duration of study, and the timing of events with respect to dates of survey.

35. The field study differed from the previous laboratory research in the following significant respects:

- the presence of tides causing the water level to vary continuously from below to about three quarters of the way up the panels

- the variability of the wave events

- the sequential action of waves at different elevations and with differing heights

- the very large number of waves involved

- the ratio of filter-size to riprap size

- the thickness of the panels ($2D_{50}$ in the laboratory work: but considerably less than that in Panels 2, 3 and 4 in the field trials)

- methods of placing the riprap, involving some degree of segregation in field conditions

- method of measuring stone size, and the grading and shape of the riprap

- the scales of the laboratory and site situations and the consequential scale effects.

Comparison of results with laboratory predictions
36. The results of the field studies were compared with the predictions from the results

contained in CIRIA Report 61 (ref. 2) and are summarised in Table 2. The general trend is that surveyed damage is rather greater than the predicted damage. It does not therefore suggest any scale effect which would result in less damage in the prototype situation.

37. A comparison was also made of the performance of the test panels in the most severe wave events with the stone sizes for stability and for failure, using alternative design methods available in the coast-protection literature (refs 1, 2, 4, 9, 10). The response of the test panels to the November 1975 and the January 1978 storms provided an empirical assessment of the stone sizes required for stability in the engineering sense of requiring little if any remedial action, and also for failure in the sense of placing the embankment at risk. These criteria were assessed somewhat subjectively, by the response of the test panels, to provide a 'best estimate' of the stone size to meet these criteria. These observational results are compared with the calculated sizes from the five available methods of design in Table 3.

38. The following conclusions may be drawn from Table 3 about the relevance of available design methods.

- Stone sizes directly deduced from field observations of performance are within the range of values obtained from design methods based on laboratory research.

- The method based on HRS research, (ref. 2) as applied here to variable wave attack gives the most consistent agreement with observations in the Wash of the stone size needed for stability.

- This method appears to underestimate the stone size for failure. The lower ratio of panel thickness to stone size used in the Wash trials, coupled with practical limitations on method of placing and resultant bulk density of the slope protection layer, probably accounts for this.

- The difference between stone size assessed for stability and for failure is less than that given by laboratory methods. It follows that a relatively small underestimate of the stone size required for 'limited damage' could lead to total failure, and emphasises the caution expressed in ref. 2 against designing for too high a level of damage.

Summary of further studies
39. Because conditions in the field tests inevitably differed from those in previous generalised laboratory tests, definitive conclusions on the scale effects could be reached only on the basis of further model tests closely reproducing both the construction of the test panels and the wave climate that caused the principle damage to them. This series of retrospective tests was carried out by HRS as part of their basic research programme and is reported

in ref. 11. Those tests further strengthened the conclusion regarding scale effects, but also drew attention to uncertainties in model testing arising from the scatter in the results of repeat tests.

40. The conclusions to be drawn on scale effects and other aspects of the research results from the field and laboratory have been further reviewed in ref. 12. One of the difficulties that emerged is the interpretation of wave observations, as there were appreciable differences between methods of analysis. The Tucker-Draper method of analysis (ref. 7) had originally been used in the field studies but some records were re-analysed by spectral methods, and this gave values for significant wave height averaging some 20% higher. It is thought that this arises from the assumption implicit in the usual spectral methods that wave heights have a Rayleigh distribution with narrow band width, giving $H_s = 1.416 H_{rms}$ and $H_{rms} = 2 \sqrt{2} \sqrt{m_o}$ where m_o is the area under the energy-frequency curve. A counting analysis to establish the average height of the highest third waves showed that $H_s \simeq 1.34 H_{rms}$; and that the Tucker-Draper method was valid but subject to errors up to ± 15% in individual events. The assessment of wave height thus presents various uncertainties, especially in shallow and intermediate water depths.

RECOMMENDATIONS FOR DESIGN

41. Unless model tests are specifically carried out for a project, riprap and filter layers should be designed on the basis of the method proposed in CIRIA Report 61 (ref. 2). No allowance for scale effects, justifying smaller sizes, should be made.

42. Designers should recognise that the behaviour of riprap has random variations, and that damage cannot be predicted accurately.

43. Designers should also be aware of the practical problems associated with placing and grading riprap which are described in CIRIA Technical Note 101 (ref. 5).

44. Attention is drawn to the problems of assessing wave climate and to the fact that errors arise from forecasting and analysing. In particular, the different results obtained from spectral and other methods of wave analysis should be considered.

45. Laying of riprap presents serious problems of quality control over grading (including segregation), median size, and layer thickness. Possible differences from specification should be taken into account at the design stage.

ACKNOWLEDGEMENT
The field studies described in this paper were funded through the Construction Industry Research and Information Association, London (CIRIA). Their permission to publish these results is gratefully acknowledged. The Authors are also very grateful to R.M. Shuttler of the Hydraulics Research Station Ltd., Wallingford, for his help. The support of many other authorities and individuals has been much appreciated.

REFERENCES
1. BURGESS, J.S. and HICKS, P.H. Riprap protection for slopes under wave attack. Civil Engineering Research Association, Research Report 4, May 1966.
2. THOMPSON, D.M. and SHUTTLER, R.M. Design of riprap slope protection against wind-wave attack CIRIA Report 61, December 1976.
3. Riprap design for wind-wave attack: a laboratory study in random waves. Hydraulics Research Station, Wallingford, Report No. EX707, September 1975.

4. THOMSEN, A.L., WOHLT, P.E. and HARRISON, A.S. Riprap stability on earth embankments tested in large and small scale wave tanks. U.S. Army Corps of Engineers, Coastal Engineering Research Centre, Tech. Memo 37, June 1972.

5. YOUNG, R.M., PITT, J.D., ACKERS, P. and THOMPSON, D.M. Riprap design for wind wave attack: long term observations on the offshore bank in the Wash, CIRIA Technical Note 101, July 1980.

6. TUCKER, M.J. Analysis of sea wave records. Proc. Instn. Civ. Engrs, Vo. 26, 1963.

7. DRAPER, L. The analysis and presentation of wave data. Proc. 10th Int. Conf. on Coastal Engineering, 1966.

8. HASSELMANN, K. et al. A paramatric wave prediction model. J. Phys. Oceanography 1976, 6(2).

9. Shore protection manual. U.S. Army Corps of Engineers, Coastal Engineering Research Centre, 1973.

10. AHRENS, J.P. Large wave tank tests of riprap stability. U.S. Army Corps of Engineers, Coastal Engineering Research Centre, Tech. Memo 51, May 1975.

11. Riprap design for wind wave attack: retrospective model tests of the measured damage to riprap panels on the offshore bank in the Wash, HRS Report ·IT213, Dec. 1981.

12. PITT, J.D., and ACKERS, P. Review of field and laboratory tests on riprap, CIRIA Report 94, 1982.

8 The core and underlayers of a rubble-mound structure

T. S. HEDGES, MEng, MICE, University of Liverpool

A review is given of the literature relevant to the design of the core and underlayers of a rubble-mound structure. The review is presented with reference to the various functions performed by the core and underlayers. Finally, a list of questions is provided to help guide the design of these elements of the structure.

INTRODUCTION

1. A rubble-mound structure (Fig. 1) generally consists of at least three main elements: a core of small stones, often 'run-of-quarry'; an armour layer of large stones or specially-shaped concrete units; and one or more inter-mediate layers - 'underlayers' - which separate the core from the armour. In addition there may be a concrete cap or wave wall. However, until the relatively recent publications of Bruun and others (refs 1, 2, 3), a casual reader of the literature on rubble-mound break-waters could surely have been forgiven for believing that the only one of these elements which was worthy of study was the armour layer. Even in the 1977 edition of the U.S. Army's Shore Protection Manual (ref. 4) there are more than twenty-two pages devoted to the armour layer whilst less than five pages are reserved for advice on underlayers and core. Yet the core and underlayers of a rubble-mound break-water have a number of very important functions to perform.

2. Included amongst the demands on the core are that:

i) it should form a satisfactory foundation for the underlayers and armour, and for any wave wall which may be required;

ii) it should provide a relatively imper-meable barrier to the transmission of wave energy;

iii) it should form a suitable working plat-form from which underlayers, armour and wave wall may be constructed (unless, of course, construction is to take place from the sea); and

iv) it should constitute a substantial portion of the total volume of the structure since it is composed of material which is relatively cheap in comparison with underlayers and armour (and which, in many cases, is available as a by-product of the quarrying for underlayer and armour stone).

3. Similarly, the underlayers have also to satisfy a number of specific demands:

i) they should provide a satisfactory foundation for the armour layer;

ii) they should act as filters to prevent core material from washing out through the voids in the armour layer; and

iii) they should act as temporary protection to the core before the armour has been placed.

4. It is the purpose of this paper to review what relevant literature exists on the design of core and underlayers. Where appropriate, the review has been carried out under headings which reflect the various demands on the core and underlayers listed above.

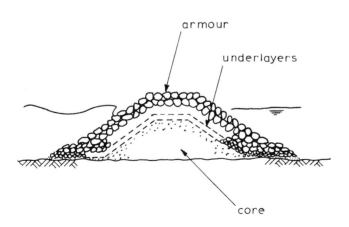

FIGURE 1: Typical cross-section of a rubble-mound structure

Breakwaters—design and construction. Thomas Telford Ltd, London, 1984

FRICTION BETWEEN ARMOUR AND UNDERLAYERS

5.   The stability of an armour unit depends not only on its own characteristics but also upon its relationship with other units. This fact is one of the reasons why there is a wide variety of shapes for concrete armour blocks. In the development of these shapes, attempts have often been made to achieve high levels of interlocking or inter-block friction. At the same time, there is a need to ensure adequate resistance to movement between the armour and the underlayers, otherwise sections of the armour layer may fail by slipping down the face of the structure.

6.   Bruun and Johannesson (ref. 1) have used some tests by Miller and Byrne (ref. 5) to indicate the influence of the size of the stones in the (top) underlayer on the angle of repose, ϕ, of the stones in the armour layer. The relationship is of the form

$$\phi = a\left[\frac{D(\text{armour})}{D(\text{underlayer})}\right]^{-0.3} \qquad (1)$$

D(armour) and D(underlayer) are the characteristic dimensions of the stones forming the armour and underlayer, respectively. The value of the constant, a, varies from 50 for spheres to 70 for tests using crushed quartzite.

7.   Based on Eqn. 1, Fig. 2 shows the variation in ϕ as a function of the ratio of the weights of stones in the underlayer to the weights of the armour stones. For loose cohesionless materials, the angle of internal friction is approximately equal to the angle of repose and so it has also been possible to indicate the variation in the coefficient of

friction, μ_f, between armour and underlayer. It is clear from the figure that an increase in the size of the material in the underlayer should bring about an improvement in the stability of the armour against sliding. The effect of stone shape is also apparent.

8.   The ability of armour to remain in place under the action of waves may be characterised in terms of the stability number, N_s:

$$N_s = \frac{H}{(\frac{\rho_s}{\rho} - 1)(\frac{W}{\rho_s g})^{1/3}} \qquad (2)$$

In this expression, H is the wave height for which armour units of weight W and density ρ_s are just stable; ρ is water density and g is gravitational acceleration. The influence of the armour slope, α, on the value of N_s has been investigated in many studies including those of Iribarren (ref. 6) and Hudson (ref. 7). They proposed the following relationships:

Iribarren: $$N_s = \frac{(\mu_f \cos \alpha - \sin \alpha)}{K^{1/3}} \qquad (3)$$

Hudson: $$N_s = (K_D \cot \alpha)^{1/3} \qquad (4)$$

K and K_D are armour stability coefficients. Note that Eqn. 3 correctly predicts that N_s will be zero when the armour slope is at its angle of repose (for which $\mu_f = \tan \phi = \tan \alpha$). Hudson's relationship, Eqn. 4, gives N_s as zero only when α is 90° (assuming that K_D is not zero).

9.   As an alternative to providing a very coarse underlayer to improve armour stability, 'binders' could be used (ref. 3) in much the same way that shear keys are employed in concrete work (Fig. 3). However, no matter what provision is made, such measures are only likely to improve stability if sliding between armour and underlayers is the critical failure mechanism. It is likely to be most important for high armour-slope angles.

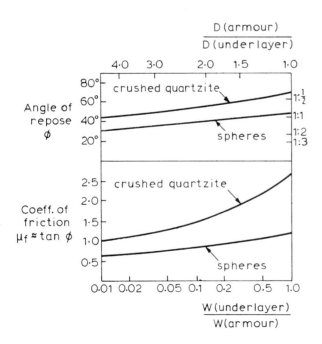

FIGURE 2:   Variation of ϕ and μ_f for armour with size of underlayer stones

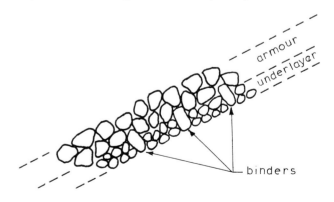

FIGURE 3:   Binders connecting armour to underlayer

Model tests by Carver and Davidson (ref. 8) in which they used stones in the top under-layer of one-fifth, one-tenth and one-twentieth of the weights of the overlying dolos units, showed no significant variation in armour stability as a result of the different underlayers. In these tests, armour slopes of 1:2 and 1:3 were used. Failure was caused, presumably, for some reason other than lack of frictional resistance on the armour by the underlayer.

EFFECTS OF MOUND PERMEABILITY ON STABILITY
10. The coefficient of permeability of a soil is the rate of flow of water per unit area of soil when there is unit hydraulic gradient. Its value depends primarily on the average size of the soil pores. In general, the finer the soil grains, the smaller the pores and the lower the permeability. The presence of even a small percentage of fines in a coarse-grained soil results in a sig-nificant lowering of the coefficient of permeability.

11. In conducting model tests on rubble-mound structures it is impossible to satisfy simultaneously the Froude and Reynolds criteria for similarity. Since tests involving gravity waves scale in accordance with the Froude number, parameters which are dependent upon the Reynolds number will be subject to scale effects. The permeability of the rubble mound is one such parameter. This fact is only important if the permeability has a sig-nificant influence on performance. Certainly permeability has a pronounced effect on the transmission of wave energy through the structure and there will be some influence on wave run-up and run-down on the face of the mound which, in turn, may affect armour stability. For these reasons it is sometimes necessary to exaggerate the size of the material in the model core. Hudson (ref. 9) and Kogami (ref. 10) describe methods for deciding on the degree of distortion.

12. Bruun and Johannesson (ref. 1) report tests in which an armour layer of 60mm granite stones rested on an underlayer of 22.5mm to 30mm stones. Three different gradings of core material were used (22.5mm - 30mm, 6.1mm - 11.5mm and 3.2mm - 6.1mm) and an impermeable slope replaced the core material as a fourth option. The armour slope was 1:1.5. The tests confirmed the influence of the core by showing a decrease in stability as the permeability was reduced. These findings were similar to those reported earlier by Hedar (ref. 11).

13. The armour on the back slope of a rubble-mound breakwater obviously must be designed to withstand the wave activity on the lee-side of the structure. The possibility of it failing as a result of wave action on the front slope must also be considered (refs 2, 12). Mound permeability may have an important influence here. For example, if overtopping occurs then the volume of water reaching the back slope will depend upon the volume which is lost through percolation into the breakwater crest (see Fig. 4(a)). Perhaps more importantly, even when there is no overtopping, armour units may be dislodged from the back slope by water and air forced through the voids in the break-water as waves impinge on the front slope. This effect may be intensified by the presence of a cap or wall (Fig. 4(b)) unless the through-flow is restricted by taking the seaward side of the cap down to the relatively imper-meable core. This would not only improve back-slope stability but would also help to keep the wave-induced uplift pressure on the cap as low as possible. Vent holes could be provided in the cap to relieve any uplift pressures which did occur.

14. Finally, in considering permeability, it is important to note that, in addition to influencing the stability of the armour, the characteristics of the core material are of primary significance to the overall stability of the rubble mound.

15. The shear strength of cohesionless soil, s, is directly proportional to the effective stress to which it is subjected (the total applied normal stress, σ, minus the pore pressure, u):

$$s = (\sigma - u)\tan \phi \qquad (5)$$

(a)

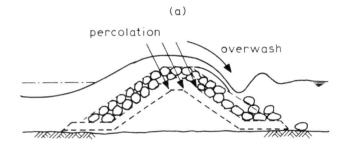

(b)

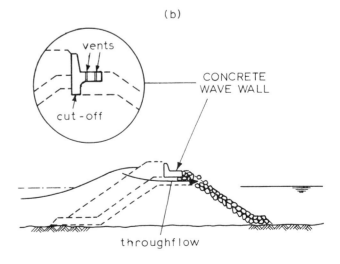

FIGURE 4:   Problems of back-slope stability

Clearly, the shear strength of the soil reduces as u approaches σ and this will happen, for example, during a rapid drawdown in water level such as occurs when a wave recedes down the front face of a breakwater (Fig. 5). The stability of the front slope should be assessed for this condition.

16. When the pore pressure equals the total applied normal stress then the shear strength reduces to zero. According to Lambe and Whitman (ref. 13), there are two common situations in soil mechanics in which this equality arises:

i)   where there is an upward flow of such magnitude that the total upward water force equals the total soil weight; the amount of water required to maintain this condition increases as the permeability of the soil increases; and

ii) where a shock is given to a loosely-deposited, fully-saturated soil which causes a volume decrease in the soil skeleton with the result that the effective stress is transferred to pore pressure.

Whilst neither of these conditions may occur during the action of storm waves on a rubble-mound breakwater, the fact still remains that the shear strength of the structure could be reduced sufficiently to cause it to fail under the repeated high impact loadings from the waves (ref. 14).

17. The above discussion on the influence of the core material upon the overall stability of the rubble mound suggests that a densely packed but fairly permeable material should be employed. Unfortunately, dumping cohesionless material through water tends to produce a loose state of deposition - a state which probably is not correctly reproduced in model tests - though the density can be improved by avoiding the use of highly angular material and adopting a wide range of stone sizes (ref. 13). To achieve a fairly permeable core, a limit may be placed on the minimum size of the material used.

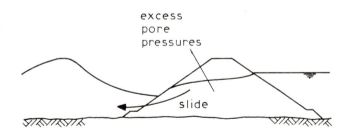

FIGURE 5:   Rubble mound subject to rapid drawdown condition

18. If a fairly permeable core material is employed in a breakwater then there will be implications with regard to wave transmission. This topic is considered in the next section.

WAVE TRANSMISSION THROUGH RUBBLE MOUNDS
19. The function of a breakwater is to protect a harbour or some part of the shore from wave attack. Consequently, during the design of the structure, consideration must be given to the amount of wave energy which will be transmitted past it. Besides the energy which passes around its ends, there may also be overtopping of the structure and transmission through it resulting from its permeability.

20. The wave transmission coefficient, K_T, defined as the ratio of the transmitted wave height to the incident wave height, can be quite large for real breakwaters. For example, Calhoun (ref. 15) has recorded K_T values of up to 40% resulting from low swell passing directly through the pores of the rubble-mound breakwater at Monterey Harbor, California. Kogami (ref. 10) shows K_T values for structures in Japan of up to 80%. In general, the transmission coefficient increases as the mound permeability increases and as the incident wave steepness reduces. In this connection it should be noted that wave grouping causes fluctuations in mean-water-level which are of very low steepness (ref. 16). Rubble mounds may transmit this low frequency beat with little attenuation whilst reducing the heights of the individual wind waves quite effectively.

21. There have been many physical model studies of wave transmission through permeable structures (refs. 17, 18) including investigations into the problem of scale effects (refs. 19, 20). Mathematical models to predict wave transmission are available also (ref. 21). As might be expected, experiments show that there is a reduction in K_T as the width of the structure, relative to the incident wavelength, is increased.

UNDERLAYERS AS FILTERS
22. Many failures of rubble-mound structures have been attributed to internal erosion. Consequently, the underlayers must be designed to act as filters to prevent core material from being washed out of the structure. The grading of the underlayers will depend upon the grain size distribution of the core material and on the maximum size of voids in the armour layer, as well as on the need to ensure sufficient frictional resistance between the armour and underlayers. Lee (ref. 22) has suggested the following criteria:

i) $\dfrac{D_{15}(\text{underlayer})}{D_{85}(\text{core})} < 5$

ii) $4 < \dfrac{D_{15}(\text{underlayer})}{D_{15}(\text{core})} < 20$

iii) $\dfrac{D_{50}(\text{underlayer})}{D_{50}(\text{core})} < 25$

iv) $\dfrac{D_{85}(\text{underlayer})}{D_{voids}(\text{armour})} > 2$

where D_{15}, D_{50} and D_{85} are defined such that 15%, 50% and 85%, respectively, of the weight of a sample, are of finer sizes. D_{voids} is the maximum size of voids in the armour layer and must be estimated or measured.

23.   Criterion (iv) applies in cases where the armour is of fairly uniform size. For rip-rap (graded armour stone) criteria (i) to (iii) should also be used to give the required relationship between the armour and the underlayer sizes. When armour units are large or when very fine core material is used it may be necessary to design a multi-layer filter system. The core would be protected by the first underlayer designed using criteria (i) to (iii). This layer would be protected, in turn, by a second, the layers continuing until the top one was sufficiently coarse that it would not wash through the voids in the armouring. In practice, the top underlayer beneath uniformly-sized armour is often composed of two layers of stones each weighing between about one-fifth and one-tenth of the weight of an armour unit. A relatively coarse top underlayer such as this, provides good resistance against sliding of the armour (see Fig. 2) and, during construction, gives reasonable protection against wave action before the armour is placed.

24.   The minimum thickness for each of the underlayers will depend upon the grading of the stones and the conditions under which they will be placed. Large, uniform stones should be at least two layers thick. Posey (ref. 23) has recommended four times the D_{85} size for graded filters laid underwater; this recommendation may be appropriate for the finer underlayers. Whatever thickness is decided upon, it should be recognised that construction will often take place under difficult conditions and sensible tolerances should be allowed.

25.   The above discussion has related to the need to provide filters between the core of the rubble-mound breakwater and its armour layer. In addition, it may be necessary to introduce a filter between the sea-bed and the structure, and between the structure and any fine material deposited as fill against its lee side (see Fig. 6). The possibility of scour at the toes of the armour slopes may also demand the provision of aprons.

SELECTION OF STONE FOR CORE AND UNDERLAYERS
26.   The selection of suitable construction materials is one of the earliest and most important matters which the civil engineering designer must consider, and particular attention has to be paid to this matter when dealing with breakwaters as the sea is amongst the most hostile of all the environments in which structures are sited. Any weakness, either in the material itself, or in the way it is employed, may be rapidly exploited by the sea.

27.   It is the aggressiveness of the maritime environment which makes the durability of a potential construction material one of the primary properties to be considered. Durability is the ability to resist wear, decay and other destructive processes. A durable material is able to withstand such processes whether they are of physical, chemical or biological origin and whether they arise within the material itself or act externally.

28.   Despite the extensive use of stone in maritime structures, little had been written about the geological aspects of its use until the recent paper by Fookes and Poole (ref. 24). They provide a wide-ranging review of the subject together with recommendations on the investigation and specification of stone for breakwaters. On the subject of stone shape, they suggest that the stone in the underlayers should be prismoidal with a maximum dimension not greater than twice the minimum dimension. For the core, the maximum dimension should be not greater than two-and-a-half times the minimum; if possible, the use of angular material in the core should be avoided. The latter requirement would help in the achievement of a high relative density (ref. 13) for the core material; employing long or flat stones could result in the formation of large voids which might lead to substantial settlement under storm conditions.

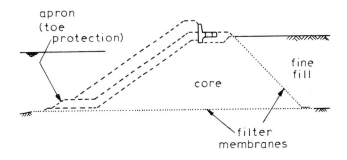

FIGURE 6:   Possible uses of filter membranes in breakwater construction

USE OF GEOTEXTILES

29. The term 'geotextile' covers all types of fabric designed to improve soil performance, including filter membranes. They have found an increasing number of uses in civil engineering over the past twenty years; Agerschou (ref. 25), Barrett (ref. 26) and others (refs. 22, 27, 28, 29) describe some of their uses in coastal work. At present, the application of geotextiles to rubble-mound breakwater construction is probably limited to acting as the filter between the sea-bed and the structure and to preventing fine material deposited as fill on the lee side of the breakwater from being washed through the core (Fig. 6). However, membranes frequently have been used also as filters beneath the armouring of sea-defence embankments and sea-walls. In these cases the underlying material is usually much finer than the core material of rubble-mound breakwaters.

30. Filter membranes have a number of significant advantages over graded filters: they have a tensile strength, a filtering ability which is largely factory-controlled and they are easily inspected. However, care must be taken in selecting these materials: for example, there may be problems with tearing and from damage by abrasion. Rankilor (ref. 30) gives details of over 130 membranes from 32 different manufacturers worldwide and describes tests for helping assess the suitability of geotextiles for their various applications (see also ref. 31).

CONSTRUCTION AND OTHER CONSIDERATIONS

31. The method used in constructing a rubble-mound breakwater, together with the need to maintain it, have important influences on its design. For example, the width of the top of the core may be wholly determined by the space required for the operation of construction plant. The sizes and numbers of the various pieces of equipment may, in turn, be affected by the production rates demanded by the construction programme.

32. When large quantities of stone are to be used in a project then it may be economically worthwhile to work the quarries in such a way that maximum use is made of their total output without need for excessive sorting or storage. In such cases it may be necessary to carry out blasting trials in order to determine the properties of the different sized stones which will be available from the quarries. The results of these trials can then be taken into account when finalising the design of the core and underlayers.

CONCLUDING REMARKS

33. The design of the core and underlayers of a rubble-mound breakwater is a complex task. To help in this work, the designer might ask himself a number of questions:

i) What range of stone sizes and shapes will be available from the quarries?

ii) Is the stone suitable in other respects, especially with regard to durability?

iii) Will the shear strength of the core be adequate?

iv) Could the relative density of the core be improved by modifying the choice of material?

v) Should a limit be placed on the minimum size of material in the core?

vi) Is the core properly represented in the physical model study?

vii) Have the underlayers been designed to act as filters?

viii) Is the top underlayer sufficiently coarse or should 'binders' be provided?

ix) How thick should the underlayers be?

x) Are the underlayers adequate as temporary armouring during construction?

xi) Have sensible tolerances been allowed for placing?

xii) How much wave transmission will there be through the structure?

xiii) Have the problems of back-slope stability been considered?

xiv) Is there sufficient scour protection at the toes of the slopes?

xv) Would the use of geotextiles be appropriate?

xvi) What can be done to facilitate construction?

34. Whilst he may not be able to give very precise answers to each of the questions (and any others which he may think proper), the designer should satisfy himself that at least each one has been considered before he finalises his design. *He should remember that the armour is designed to prevent erosion of the structure's surface, not to act as a retaining 'skin' against deep-seated failure brought on by poor design of the core and underlayers.* This paper will have served its purpose if it encourages just a little more thought to be given to such matters.

ACKNOWLEDGEMENTS

35. I am grateful to my colleagues, Mr. G. J. W. King and Dr. E. A. Dickin, for commenting on a draft of this paper.

REFERENCES

1.   BRUUN, P. and JOHANNESON, P. Parameters affecting stability of rubble mounds. J. Waterways, Harbors and Coastal Engng. Div., Am. Soc. Civ. Engrs., Vol. 102, No. WW2, May, 1976, 141-164.

2.   BRUUN, P. Common reasons for damage or breakdown of mound breakwaters. Coastal Engineering, Vol. 2, 1979, 261-273.

3.   BRUUN, P. and KJELSTRUP, Sv. Practical views on the design and construction of mound breakwaters. Coastal Engineering, Vol. 5, 1981, 171-192.

4.   U.S. ARMY COASTAL ENGINEERING RESEARCH CENTER. Shore Protection Manual 3rd Edition. U.S. Govt. Printing Office, Washington, 1977.

5.   MILLER, R.L. and BYRNE, R.J. The angle of repose of a single grain on a fixed rough bed. Tech. Rep. No. 4, Dept. Geophys. Sc., University of Chicago, May, 1965.

6.   IRIBARREN CAVANILLES, R. Una formula para el calcula de los diques de escollera. Revista de Obras Publicas, Madrid, 1938.

7.   HUDSON, R.Y. Laboratory investigations of rubble-mound breakwaters. J. Waterways and Harbors Div., Am. Soc. Civ. Engrs., Vol. 85, No. WW3, Sept., 1959, 93-121.

8.   CARVER, R.D. and DAVIDSON, D.D. Dolos-armoured breakwaters: special considerations. Proc. 16th Coastal Engng. Conf., Am. Soc. Civ. Engrs., 1978, 2263-2284.

9.   HUDSON, R.Y. Coastal Harbors. In Coastal Hydraulic Models by R.Y. Hudson et al., Special Rep. No. 5 for U.S. Army Coastal Engng. Res. Center, U.S. Govt. Printing Office, Washington, 1979, 202-283.

10.   KOGAMI, Y. Researches on stability of rubble-mound breakwaters. Coastal Engng. in Japan, Vol. 21, 1978, 75-93.

11.   HEDAR, P.A. Stability of rock-fill breakwaters. Akademiförlaget-Gumperts, Göteborg, 1960.

12.   WALKER, J.R., PALMER, R.Q. and DUNHAM, J.W. Breakwater back slope stability. Proc. Civil Engng. in the Oceans III, Am. Soc. Civ. Engrs., 1975, 879-898.

13.   LAMBE, T.W. and WHITMAN, R.V. Soil Mechanics, S.I. version. Wiley, New York, 1979.

14.   HARLOW, E.H. Large rubble-mound break-water failures. J. Waterway, Port, Coastal and Ocean Div., Am. Soc. Civ. Engrs., Vol. 106, No. WW2, May, 1980, 275-278.

15.   CALHOUN, R.J. Field study of wave transmission through a rubble mound break-water. M.S. thesis, U.S. Navy Postgraduate School, Monterey, March, 1971.

16.   LONGUET-HIGGINS, M.S. and STEWART, R.W. Radiation stresses in water waves: a physical discussion with applications. Deep-Sea Res., Vol. 11, 1964, 529-562.

17.   SOLLITT, C.K. and CROSS, R.H. Wave transmission through permeable breakwaters. Proc. 13th Coastal Engng. Conf., Am. Soc. Civ. Engrs., 1972, 1827-1846.

18.   KONDO, H. and TOMA, S. Reflection and transmission for a porous structure. Proc. 13th Coastal Engng. Conf., Am. Soc. Civ. Engrs., 1972, 1847-1866.

19.   DELMONTE, R.C. Scale effects of wave transmission through permeable structures. Proc. 13th Coastal Engng. Conf., Am. Soc. Civ. Engrs., 1972, 1867-1872.

20.   WILSON, K.W. and CROSS, R.H. Scale effects in rubble-mound breakwaters. Proc. 13th Coastal Engng. Conf., Am. Soc. Civ. Engrs., 1972, 1873-1884.

21.   MADSEN, O.M., SHUSANG, P. and HANSON, S.A. Wave transmission through trapezoidal breakwaters. Proc. 16th Coastal Engng. Conf., Am. Soc. Civ. Engrs., 1978, 2140-2152.

22.   LEE, T.T. Design of filter system for rubble-mound structures. Proc. 13th Coastal Engng. Conf., Am. Soc. Civ. Engrs., 1972, 1917-1933.

23.   POSEY, C.J. Protection of offshore structures against underscour. J. Hydraulics Div., Am. Soc. Civ. Engrs., Vol. 97, No. HY7, July, 1971, 1011-1016.

24.   FOOKES, P.G. and POOLE, A.B. Some preliminary considerations on the selection and durability of rock and concrete materials for breakwaters and coastal protection works. Q.J. Eng. Geol., Vol. 14, 1981, 97-128.

25.   AGERSCHOU, H.A. Synthetic material filters in coastal protection. J. Waterways and Harbors Div., Am. Soc. Civ. Engrs., Vol. 87, No. WW1, Feb., 1961, 111-123.

26.   BARRETT, R.J. Use of plastic filters in coastal structures. Proc. 10th Coastal Engng. Conf., Am. Soc. Civ. Engrs., 1966, 1048-1067.

27.   DUNHAM, J.W. and BARRETT, R.J. Woven plastic cloth filters for stone seawalls. J. Waterways, Harbors and Coastal Engng. Div., Am. Soc. Civ. Engrs., Vol. 100, No. WW1, 1974, 13-22.

28.   WELSH, J.P. and KOERNER, R.M.
Innovative uses of synthetic fabrics in coastal
construction. Coastal Structures 79, Am. Soc.
Civ. Engrs., 1979, 364-372.

29.   RANKILOR, P.R. The design of a two-
layer membrane/webbing filter system for a
marine causeway wave defence system in the
Gulf of Arabia. Q.J. Eng. Geol., Vol. 15,
1982, 227-231.

30.   RANKILOR, P.R. Membranes in ground
engineering. Wiley, Chichester, 1981.

31.   I.C.I. FIBRES. A guide to test
procedures used in the evaluation of civil
engineering fabrics. I.C.I. Fibres,
Pontypool, 1981.

9 The design of armour systems for the protection of rubble mound breakwaters

W. F. BAIRD and **K. R. HALL**, W. F. Baird & Associates, Coastal Engineers Ltd, Ottawa

A summary of the various forms of armour systems required to protect rubble mound breakwaters from wave attack that have been employed since the early 1800's, is presented. The design and performance of these protection systems are reviewed. It is noted that design procedures published in recent years have severe limitations and that their use may have led to damage of many breakwaters. Improved or alternative design procedures for armour protection systems that overcome some of these limitations, are outlined.

1 INTRODUCTION

Wave protection systems, or armour layers, evolved as breakwaters consisting of dumped stones or quarry run, were built in deeper water and at locations exposed to increasingly severe environmental loadings.

The first rubble mound breakwaters simply consisted of quarry run dumped at a location where protection from wave action could be given to a harbour. Wave action acting on the exposed face of the breakwater resulted in reshaping of the breakwater profile into a flatter slope and natural armouring of the outer layers due to removal of the finer material. These breakwaters were often breached and required frequent maintenance.

It was a logical development that large stones, cut or blasted from suitable quarries, were placed on the exposed face of the breakwater in an attempt to reduce maintenance and eliminate breaching of the breakwater during very severe storms. In the early designs, armour stones were only placed on the crest of the structure, extending to the low water line, because of the lack of suitable construction equipment.

At some locations, a lack of a suitable quarry or a physical limit to the size of the armour stone that could be produced, transported and placed, resulted in a breakwater that was still damaged by severe wave action and required extensive repair to remain functional. During the early 19th century advances in the development of concrete for marine applications enabled several designers to use larger, more massive "artificial" armour units. As early as 1834, Reference 1, concrete armour units, rectangular in cross section, weighing in the order of 10 to 60 tons were placed in either a regular or random fashion on a number of breakwaters. By the early 1900's, concrete blocks were being extensively used for breakwater protection throughout the world.

More recently, as breakwaters have been required at more exposed locations and in deeper water, a variety of complex shapes of concrete armour units have been developed. These units rely on a certain degree of interlocking with adjacent units, thereby reducing the block weight required for stability against wave action during storms.

The engineer responsible for the design of a breakwater has not had available to him complete and proven design procedures that encompass all aspects of the project. Consequently damage to breakwater armour systems has been observed relatively frequently in recent years.

In the past, when designs were based only on previous experience, damage or complete failure of the protection system was common. However, the essential requirement to exchange information or prototype experience was appreciated. The performance of breakwaters in the late 1800's was, for example, extensively discussed at two engineering conferences, the International Engineering Conference in St. Louis in 1904 and the International Navigation Congress in Milan in 1905. Observations were objectively presented in order that errors would not be repeated. Two interesting examples were the damage caused by the 1898 storm on the Genoa breakwater and the 1893 storm on the Bilbao breakwater. At Genoa in February 1898, 25-foot waves caused 40-ton concrete blocks to be displaced distances of over 160 feet, and at Bilbao in 1893, the armour layer of 60-ton concrete blocks was almost completely removed.

In recent years, the severe damage to the massive Sines breakwater has initiated a number of investigations and reviews of design procedures that will be beneficial to future designs. In addition, many other breakwaters including San Ciprian, Bilboa, Kahului, Diablo Canyon, Baie Comeau, Riviere au Renard and Gansbaai, have suffered severe damage that required extensive

Breakwaters—design and construction. Thomas Telford Ltd, London, 1984

107

maintenance.

The objective of this paper is to review and summarize existing and past procedures for designing armour systems required to protect rubble mound breakwaters from wave attack. This review covers armour layers consisting of quarried stones either randomly or regularly placed in the armour layer, and concrete armour units placed randomly or regularly on the breakwater.

It is noted that published design procedures that are currently used for some designs, have severe limitations that may produce inefficient and costly designs, as well as designs that may be unstable during the design storm. The limitations of these procedures are described and an alternative procedure is outlined. However, the experience gained through the exchange of information describing the performance of prototype breakwaters, and by inspection of successful and damaged structures, will remain an essential ingredient to the development of a successful design.

2 DESIGN OF PROTECTION SYSTEMS USING QUARRIED STONE

2.1 Introduction

Early designs of rubble mound breakwaters, for which detailed technical descriptions are available, did not include any form of armour layer. The breakwaters consisted of dumped quarry run and the main consideration in the design was the slope of the exposed face of the breakwater, Figure 1. Technical literature from the 1800's and early 1900's, References 2 and 3, report many observations of profiles of these structures following storm action without any discussion of the requirement for armour stones. Typically, slopes of between 4 to 1 and 7 to 1, were measured in the vicinity of the water level.

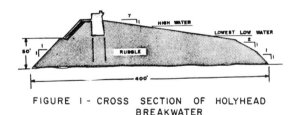

FIGURE 1 - CROSS SECTION OF HOLYHEAD BREAKWATER

John Rennie's original design for the Plymouth breakwaters is a classical example of this type of breakwater, Reference 4. The designed cross section was based on observation of the profile of other breakwaters, (in particular, the Cherburg breakwater), developed during periods of wave action, Figure 2.

It was appreciated by the design engineers, that at exposed locations this type of breakwater would require extensive maintenance. Cunningham, Reference 5, in a 1928 treatise on harbour engineering, notes that "mounds are lacking in the quality of permanence". Rennie, Reference 4, described the large volume of

material that was washed over the top of the Plymouth breakwater and estimated this volume to be in the order of 30 per cent of the total material required for construction. Rennie later added an armour layer to his design for the Plymouth breakwater.

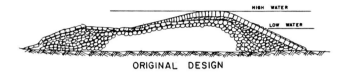

ORIGINAL DESIGN

EXISTING STRUCTURE

FIGURE 2 - PLYMOUTH BREAKWATER

Since the 1800's, many rubble mound breakwaters with armour layers of quarried stone have been built at many locations around the world. These armour layers can be classified, in general, as being constructed of randomly placed stone, normally in two layers, or regularly placed stones, normally consisting of one layer of stones placed side by side in as regular an order as possible.

2.2 Randomly Placed Quarried Stone

This breakwater concept in its classic form, consists of a core of small granular material protected by two layers of randomly placed or dumped large stones, generally uniform in size. Filter layers of selected stones are required between the core and armour layer to stop the removal of the small material through the armour layer, Figure 3.

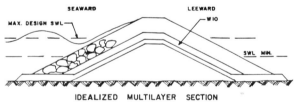

IDEALIZED MULTILAYER SECTION

FIGURE 3 - RANDOMLY PLACED ARMOUR STONE

Many empirical formulas have been proposed, following the pioneering work of Iribarren, Reference 6, in 1938, to assist in determining the weight of the stones required to resist the wave action of the design wave conditions. At the present time, the Hudson formula, Reference 7, is the most extensively used of all the formulas available. This is because of the considerable amount of laboratory data that are available to define the stability coefficient required in the formula.

It is generally recognized that design formulas should only be used for preliminary design estimates. A Permanent International Association of Navigation Congresses' (PIANC) report,

Reference 8, states "The Commission considers the present stability formulae for rubble mound structures to have significant limitations. It is only for a preliminary assessment of the dimensions of quarry stone armour units that the formulae might be applied".

A design obtained using a published formula does not represent the unique condition of the site and the local materials, such as the characteristics of the local wave climate, the yield of local quarries, and the properties of stones available from the quarry. If a design procedure does not consider the yield of the quarry, significant wastage of the quarried material may result. The volume wasted depends on the total volume of stone that must be quarried to obtain the required armour stones and whether or not this volume is in excess of the volume required for the core of the breakwaters.

All formulas use coefficients that are or have previously been determined from the results of experiments with a physical model of a breakwater. These experiments also have many limitations.

It is the authors' opinion that many engineers consider that the Hudson formula produces a conservative design for an armour layer of randomly placed stones. Perhaps one reason for the apparent success of the structures designed according to this procedure, is that most rubble mound breakwaters with armour layers of quarried stones, are built at locations that are relatively protected and where the maximum wave heights may be limited by the depth of water seaward of the site.

A recent (1976) report, Reference 8, recommends that a preliminary design be developed using a selected formula, and that this should be followed by a series of model tests of the breakwater in a wave flume to provide the final design.

The key factor in this procedure is how well the prototype structure is modelled and how well the design storm is simulated in the tests. The problems associated with model testing are discussed in Section 5.

2.3 Regularly Placed Quarried Stone

This concept has been frequently used at locations where the available armour stones are approximately rectangular in cross section. Stones are placed individually, side by side, working up from the bottom of the layer. Each stone is placed with its long axis perpendicular to the face of the breakwater, Figure 4. The long axis of the stones should be in the order of twice as long on the other dimensions. Whereas randomly placed stones may not obtain significant resistance to movement by friction forces from adjacent stones, regularly placed stones achieve much greater stability as a result of support from adjacent stones. Model studies, References 10 and 11, have demonstrated the potential for higher stability of a

regularly placed armour layer under wave attack when compared to a randomly placed stone layer.

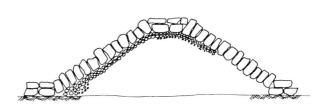

FIGURE 4- REGULARLY PLACED ARMOUR STONE

This form of breakwater has been extensively used in the Great Lakes since the early 1900's, Reference 9, and is still used extensively by Canadian engineers. It has also been used in the north-west United States. The concept has application in relatively shallow water locations where a crane operator can place each stone side by side.

Understandably, designs should be developed using a model study of the specific site conditions, because the stability of the concept is dependent on the shape of the available blocks, the resulting frictional forces achieved between blocks, and the characteristics of the armour layer resulting from the construction method selected. All of these parameters are unique to the site; therefore, generalized sets of tests would have little application.

There are limitations to the concept. If the support from adjacent stones is lost because of disintegration of a poor quality stone or because of settlement of the mound, then the unsupported stones act as they do on the randomly placed layer. The armour layer must have rigid and stable support at its base. If the support at the base or lower end of the armour layer is lost or moved, then the integrity of the armour layer is also lost. The design relies upon the additional stability, resulting from interlocking of the units, in order for the armour layer to remain intact during the design storm. If the interlock is lost because of a loss in support, then rapid unravelling of the armour layer may occur.

2.4 Alternative Design With Quarried Stones

There are alternative designs for the protection of rubble mound structures which are radically different from other breakwater concepts designed for similar environmental conditions. The following is a brief description of the development of a structure designed to maximize the yield of the available quarry and to be built with readily available land-based construction equipment.

A preliminary design for a 700m long structure extending into 17m of water was undertaken as part of an extensive runway upgrading project. Development of the wave protection scheme for the rockfill structure, was undertaken using a series of two and three dimensional hydraulic model studies. The structure was designed to survive a storm that had a 36 hour duration, a peak significant wave height of 10m and a peak wave period of 12 seconds. The final design consisted of a 25m wide outer berm of 4.5 to 19 ton angular quarry stones between the elevations of -17m and +3m and a conventional two stone layer built between +3m and +10m elevation, Figure 5.

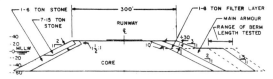

FIGURE 5-ALTERNATIVE DESIGN CONCEPT
(HALL ET AL 1983)

The stability of the structure is developed as armour stones are moved and the outer berm is reshaped by wave action. A natural armouring of the surface layer of armour stones occurs due to sorting of stones which maximizes the interlocking between stones.

During the initial periods following construction, when no natural armouring has occurred, the extent of stone movement is greatly reduced because of the capability of the porous armour layer to dissipate wave energy. Test results indicate that if design wave conditions are exceeded, catastrophic failure, (rapid disintegration of the armour layer), would not occur. Details of the design are discussed in Reference 12.

Although this type of design concept requires large volumes of armour stone, use of the yield of the local quarry can be maximized because, generally, large volumes of rockfill will be required for many projects which will automatically result in the production of large volumes of armour size material.

3 DESIGN OF PROTECTION SYSTEMS WITH CONCRETE ARMOUR UNITS

3.1 Introduction

Concrete blocks were required for breakwaters at exposed locations where the maximum sized stones available from local quarries were constantly removed by wave action.

The French engineer, Poirel, appears to have been the first to determine that concrete blocks could be manufactured for installation on a breakwater and then to successfully implement the design, Reference 1. Poirel designed 22 ton rectangular blocks (parallelopipeds) for random placement on the Algiers breakwater in 1834, Figure 6. On the crest of the same breakwater, 200 ton concrete monoliths were cast in place.

Following the first use at Algiers, similar blocks were used extensively in France including the main breakwater at Marseilles, Figure 6, which was also designed by Poirel. Other early uses of concrete blocks included 25.5 ton blocks at Cette, La Ciotat, Port Vendres, 44 ton blocks

at Cassis, 60 ton blocks at Bilbao, 30 ton blocks at Madras, 17 to 30 ton blocks at Columbo and 8 ton blocks at Osaka. These and other examples are described in the literature of the late 1800's and early 1900's, References 1, 3, 5, 13, 14 and 15.

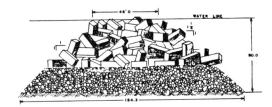

CROSS SECTION OF ALGIERS BREAKWATER

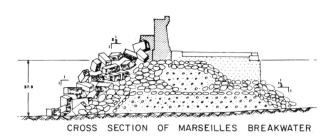

CROSS SECTION OF MARSEILLES BREAKWATER

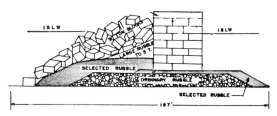

CROSS SECTION OF MADRAS BREAKWATER
FIGURE 6-EARLY EXAMPLES OF CONCRETE
BLOCK ARMOUR LAYERS

The blocks were placed either in a random fashion or in a regular, stepped order. The structures noted above contained blocks randomly placed, whereas the breakwater at Genoa built in 1888 contained 40-ton rectangular blocks placed in a stepped sequence. At the International Navigation Congress in Milan in 1905, it was concluded that "The method of depositing these blocks at random, appears the best as regards resistance and maintenance .... the method of setting the blocks in regular courses offers serious objections as the blocks are liable to be disturbed by the settlement of the rubble base and to be completely destroyed during storms".

In the 1900's, the use of the concrete blocks increased as more harbours were required in deeper water. Consequently, in the past 30 to 40 years, a number of different forms of blocks have been developed. The objective of these developments was clearly to obtain an armour unit that was lighter than the rectangular block, but exhibited the same stability under wave action. The units were intended to be easier to handle and less expensive due to savings in materials.

As was the case with stones, two general categories of armour layer construction using concrete armour units exist: designs that contain randomly placed units, and designs that contain armour units placed in a regular pattern.

3.2 Randomly Placed Concrete Units

Rectangular shaped blocks have been placed in a random fashion on breakwaters since 1834. The first extensively used concrete unit with a complex shape was the tetrapod developed in France in the 1950's. Some early applications were the use of 25 ton units at Crescent City, USA, Figure 7, Safi, Morocco and Rota, Spain. The dolos unit, developed in the 1960's and first used at East London, S.A. (17 ton units), has seen extensive use worldwide in the past two decades. Many other types of units have been developed and are described in recent technical literature. Examples include the quadrapod (Reference 16), tetrahedron (Reference 17, 18), stabit (Reference 19), toskane (Reference 20), grobbelaar (Reference 20) and akmon (Reference 21).

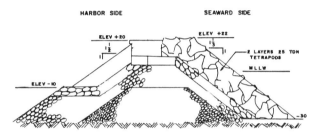

FIGURE 7 - TYPICAL TETRAPOD CROSS SECTION CRESCENT CITY

These complex shapes were designed to interlock with adjacent units in the armour layer, thereby reducing the weight of the unit required to be stable in a given wave climate. This interlocking made it difficult for wave-induced hydraulic forces to remove an individual unit without having to move a mass of units.

An increase in stability can also be attributed to the fact that for certain geometric shapes, the percentage of voids in the armour layer increases drastically compared with conventional rectangular block armour layers. As waves propagate into a porous layer, much of the wave energy is dissipated within the armour layer. This results in reduced hydrodynamic forces acting on the units at the surface of the armour layer, which are often the most vulnerable to excessive movement.

To assist the designer in selecting the weight of the required unit, the same formulas developed for quarried stones are used, but with stability coefficients adjusted to represent the performance of the various concrete units. Use of these formulas is subject to a number of limitations that are discussed in detail in Section 5.

Review of the technical literature shows that little attention has been given to the structural design of the individual armour unit.

Engineers, with few exceptions, have assumed that each unreinforced unit is capable of resisting the static loads applied by adjacent units, the forces resulting from wave action and the dynamic loads that occur when waves cause the units to rock.

There are some notable examples where reinforced units have been used, including 25 ton tetrapods at Sofi, Morocco built in 1955 and 42 ton dolosse placed at Humbolt, USA in 1971-72. However, the amount of steel placed in these units was not determined from the results of an analysis of the loads expected on the unit while in service.

There is some discussion in the technical literature concerning the optimum geometry of the dolos unit to provide improved strength characteristics. These recommendations do not result from an analysis of expected loads on the unit and the resulting stress distributions within the unit that will occur on the breakwater during wave attack.

Most major structures built in recent years using concrete armour units have been designed with the assistance of the results of model studies of the breakwater subjected to wave attack. Problems associated with model studies are discussed in Section 5.

In most of these model studies, the material used to model the units has not simulated the strength of prototype concrete and therefore, has not been able to simulate unit breakage that may occur in prototype. Consequently, any model results for a situation where breakage or significant abrasion may have occurred in prototype are invalid.

It is concluded that the design of breakwaters using concrete armour units must be based on well-designed model studies in which static and dynamic forces occurring during a storm are defined. The armour units are then structurally designed, using standard engineering procedures, to resist the defined loads. A design following this procedure has not been described in the technical literature.

3.3 Regularly Placed Concrete Units

An armour layer of regularly placed concrete armour units can be compared to the regularly placed armour stone concept described in Section 2.3. Considerable reduction in the required armour unit weight when compared to randomly placed units for given wave conditions is possible if the blocks are fitted together in a regular fashion. Examples of units designed for this type of construction and described in the technical literature include the tribar (Reference 22), Figure 8, seabee (Reference 23), cob (Reference 24), shed (Reference 25) and hollow tetrahedron (Reference 18).

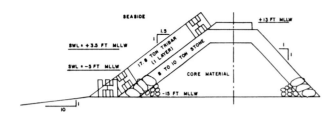

FIGURE 8 TYPICAL TRIBAR BREAKWATER
CROSS SECTION (NAWILIWILI USA)

Construction requires careful placement of each
unit so that it fits tightly against adjacent
units. This procedure requires visual observation of the placement by the crane operator or
the assistance of a diver in voice contact with
the operator. If some units are not placed
correctly and are dislodged by wave action then
that section of the armour layer will quickly
collapse because of the loss of interlocking
between the units.

If some units break, or the base support to the
armour layer moves, or if differential settlement of the mound occurs, then the loose units
behave as they would in the random placement
concept. However, during the design storm
these units will be displaced from the breakwater because they have insufficient weight to
remain on the armour layer without support from
adjacent units, and rapid unravelling of the
armour layer will occur.

The extremely high stability of this concept has
been demonstrated in model tests conducted under
ideal laboratory conditions. The main limitation appears to be in the practical considerations of providing an absolutely fixed base to
the armour layer, an even underlayer on which
the units are placed, a mound that will not
exhibit significant differential settlements,
quality control in the manufacture of the units
so that there are no faulty units, and in prototype placement procedures that ensure each
unit is correctly located without damage. In
model tests, the most important consideration
may be the simulation of the hydraulic head that
occurs across the armour layer during wave action.

4 DESIGN OF ALTERNATE BREAKWATER CONCEPTS

There is no standard breakwater design, using
either quarried stone or concrete armour units.
The design engineer should always be thinking
of new concepts that may be suitable to the site
specific conditions such as environmental loadings, physical conditions, material sources, and
methodology availability.

The main variation from breakwater concepts considered in the past, has been the inclusion of
a submerged mound or berm in front of the breakwater.

The berm or submerged mound causes waves over a
certain size to break before reaching the main
armour layer. Therefore, a limit to the

maximum wave height reaching the main structure
is imposed, which is dependent on the overall
characteristics of the berm or submerged mound.
The armour layer can be designed for this reduced wave height. The submerged mound or berm
will contain smaller armour units than are required in the standard surface piercing concept,
because it is submerged and is not subjected to
the large forces occurring at the water surface.

At some breakwaters, a bituminous compound has
been applied to the armour layer. The intended
result is a continuous and impervious layer that
is molded to the material beneath it. This form
of armour layer is not discussed in this report.

5 DISCUSSION

In the previous sections, a brief history of the
design of armour protection systems for rubble
mound breakwaters has been presented. The design procedures that have been and are currently used to design these armour protection systems were discussed.

In general, it is concluded that there are two
basic design methods which involve the use of
either hydraulic model studies or empirical
equations that relate wave height to the weight
of the armour unit. Both procedures have many
limitations. Furthermore, there are many examples of breakwaters designed using these procedures that have been severely damaged by wave
action.

In this section, these procedures and their
limitations are discussed in greater detail.
The possible causes that are reported to be responsible for the damage to some breakwater
armour layers in recent years, are reviewed.

5.1 Formula for Estimating the Weight of an
Armour Unit

A number of equations have been proposed to assist in determining the size of armour stones
for a given site, Reference 8. The first equation to be extensively used was that presented
by Iribarren, Reference 6. In the United States
Hudson and Jackson reviewed the Iribarren formula with the benefit of the results of an extensive series of model studies. They found
that the friction coefficient required in Iribarren's formula could not be defined, Reference 7, and, therefore, proposed a simpler formula that did not include a specific friction
factor. The resulting equation, which is now
extensively used, is known as the Hudson formula.

The Hudson formula is as follows:

$$W = \frac{H^3 w_r}{Kd (Sr-1)^3 \cot\theta}$$

where: W = weight of the armour unit
       H = wave height
       w_r = unit weight of stone or concrete

Sr = specific gravity of the armour unit
cot θ = slope of armour
Kd = damage coefficient

Since development of the Hudson formula, the dependence of armour stone stability on wave period has been established. Damage coefficients that are a function of the wave period have been proposed by Ahrens, Reference 26, for use in the Hudson formula.

The Hudson formula has, in recent years, been extensively used to describe the performance of interlocking concrete units or stones and concrete units that are regularly placed. However, a report presenting the Hudson formula, Reference 27, states:

"For breakwaters constructed by dumping, or by placing armour units essentially pell-mell, the forces resisting displacement are the buoyant weight of the individual units and the friction between units. Except for isolated instances where wedging is involved, friction between armour units can be neglected,...."

Therefore, the Hudson formula was derived on the basis that friction could be ignored. However, this same formula is now used to describe the performance of concrete blocks that depend primarily on interblock friction.

The damage coefficient required in the Hudson formula is derived from the result of model studies. Therefore, this coefficient is a function of the model test conditions and the interpretation of the observer.

The main limitation of the model studies and, therefore, of Hudson's formula are as follows:

- It has not been established what wave height should be used in the formula (Hmax, H(10%) or Hs, etc.) if the published coefficient was determined with regular waves.

- The dependence of the damage coefficient on wave period, wave steepness, the offshore parameter, or the type of breaker (if breaking occurs) has not been well defined.

- Published coefficients are, in general, only available for the 'standard' armour layer with direction of wave attack at right angles to the breakwater. In many cases, the design wave direction does not occur at right angles, to the structure. Furthermore, the weakest section of the breakwater may be at a transition, for example, where the size of armour unit or slope changes close to the head of the breakwater.

- Friction between units is only considered through the damage coefficient.

- The prototype breakwater may have different characteristics than the model structure that may affect the stability of the armour layer. Examples include the inclusion of a solid superstructure and the use of core material with a different permeability than in the model.

- The model breakwaters, if they were similar to the prototype structures, were built to exact dimensions and specifications that do not correctly model the results of prototype construction procedures.

- The model breakwater was built with materials that have different characteristics than the prototype materials. These characteristics may include the porosity of the core and filter layers and roughness, angularity, shape, density, durability, and strength of the armour units.

- Few tests have been completed to determine coefficients to assist with the design of the head of the breakwater.

- The published damage coefficients are dependent on the details of the model, the test procedures and how the observer reported the results. Model studies are discussed in the following section.

As previously noted, use of the Hudson formula may produce an inefficient design that does not maximize the use of the quarry, and it does not provide any assistance in utilizing a range of stone sizes for the armour layer (which may produce a more stable design).

5.2 Model Studies to Determine the Stability of the Armour Layer

Many laboratories throughout the world have undertaken model studies of breakwaters. Frequently, the results of the test program are published by providing damage coefficients for use in Hudson's formula. It is difficult for the design engineer to interpret the results of the model studies or to compare the results of different laboratories because of differences in the procedures followed in the test program. Some of these differences are as follows:

(1) Characteristics of Armour Units

The following properties may vary considerably in the different laboratory experiments:

- surface friction of units (the various materials used for the model units include concrete, sulphur and sand mixture and plastic)

- dimensions of units

- density of material used for construction of the model units (allowance should also be made if fresh water is used for a salt water location)

- strength and durability of the model material (particularly for concrete units). Ideally the tensile, compressive, and shear strengths, fracture toughness and elasticity of the unit should be modelled.

(2) Construction of the Model Breakwater

In some studies, the breakwater was built in a dry flume and in others, with water in the flume. In some studies, construction of the model has followed prototype construction practices. In other studies, individual units have been unrealistically placed with tight interlocking, obtained by carefully locating the unit and pressing the unit into the armour layer. Model construction should follow prototype placement procedures.

(3) Wave Parameters

The first model studies were conducted with sinusoidal waves of constant height and period. Later, laboratories used irregular waves produced either by the superposition of discrete sinusoidal components or by filtering a white noise signal that controls the wave generator. Many laboratories have also developed the capability of generating water surface profiles identical to surface profiles recorded in the sea. Such profiles may contain a predefined series of large waves directly following each other (wave groups).

Some laboratories have also produced waves by direct wind generation in the flume. The wind may produce a realistic wave steepness as well as a realistic distribution of heights and periods. However, generating waves of a sufficient size for breakwater tests is not practical using this method.

The wave conditions must simulate the action of the real sea during the course of a storm. This implies that breaking waves must be correctly simulated; the probable maximum wave height and groups of large waves must be modelled. Some of these phenomena may be site specific. Furthermore, variations in water level must be considered as this may change the way waves break on the structure and the area of the breakwater that is subjected to wave attack.

During a test of a breakwater, the time history of the design storm should be simulated to represent the changes in wave height and wave period that occur during the storm. The duration of a storm may be an important parameter in determining the stability of a given armour layer.

(4) Test Procedures

Tests undertaken at different laboratories have had variations in the scales at which model studies have been carried out, the duration of tests and the sequence of testing at different wave heights and periods. Furthermore, the depths of water, the tidal ranges, the angles of wave attack, and the details of the structure, such as the structure height and the extent of the armour layer, have varied. All affect the published results.

(5) Reporting of Results

A review of reports of model studies shows great variation in reporting the results of wave interaction with the armour units. The definition of damage to the armour layer is very controversial and has been described as the oscillation of individual units in the armour layer, the movement of units from one location to another in the armour layer, or the displacement of units from the armour layer. The amount of damage has been expressed either as a percentage of all units in the structure or as a percentage of units within a defined area, usually where the damage has occurred. It should be noted that oscillation, or rocking of individual units is difficult to observe because it often occurs immediately below the water surface of a breaking wave.

Damage has also been described in terms of changes in the profile or cross section of the armour layer that occurred during the test.

There has been little conformity in the reporting of model study results and, as a consequence, there are considerable variations in the reported performance of various breakwaters.

(6) Model Scales

There has been a great variation in the geometric scale at which the model breakwaters have been built. Breakwater models are designed according to the Froude criterion to correctly simulate gravitational forces, while flow in the armour and filter layers, to the extent that it affects stability, can only be modelled correctly if similarity of viscous forces is ensured by satisfying the Reynold's criterion. Simultaneous satisfaction of both criteria is physically impossible, therefore, there is the potential for significant scale effects in modelling the viscous forces unless the model scale is selected so that the Reynolds problem is negligible. Dai and Kamel, Reference 28, provide guidelines for calculating the effective Reynolds number in the model and relating this to an acceptable level that will eliminate most scale effects.

In the authors' opinion, well designed and carefully executed model studies are an essential part of the design of a large rubble mound structure located in deep water. It is absolutely essential that the above items be given careful attention.

The engineer responsible for the breakwater design must become fully involved with the test program and he may involve the contractor, inspectors, and equipment operators. The engineer must ensure that prototype conditions are represented as accurately as possible.

In particular, the responsible engineer must be present during the tests to observe the performance of the breakwater during the design storm and to interpret the observations and results.

5.3 Recent Failures of Rubble Mound Breakwaters

In recent years, armour layers of many large breakwaters have been damaged by wave action, References 29, 30, 31, 32, 33, 34, 35, 36, 37, 38, 39, 40, 41 and 42. It is important that case histories of both successful and unsuccessful structures be presented in the technical literature because there is no substitute for prototype experience. This requires that authorities responsible for breakwaters initiate and maintain well-designed procedures for the inspection and monitoring of these structures.

In general, the causes of damage to breakwater armour layers can be placed in four categories:

1. The criteria used in the design of the structure (particularly the wave criterion) were exceeded during the storm causing the damage. Other phenomena for which design criteria may have been established include earthquake accelerations, long period waves (such as tsunamis) and geotechnical considerations for the subsoils.

2. The procedures followed during the development of the design were incomplete or incorrect for the site. For example, the formula used to determine the weight of the armour unit was used incorrectly or was not appropriate, or model studies did not accurately simulate the real conditions at the site or the properties of the prototype structure.

3. The breakwater was not built according to the plans prepared by the design engineer.

4. The materials used for the construction were substandard.

Some other specific causes that are considered to have contributed to recent failures are described below:

5. Design Wave Underestimated

   The design wave or its return period used in the design process was underestimated because of an incomplete knowledge of the wave climate or because of an incomplete analysis of the influence of the bottom topography on increasing the heights of incident waves on the structure.

6. Breakage of Concrete Armour Units

   Concrete armour units moved and then broke because the units had not been structurally designed for the loads acting on individual units. In the model studies, the units did not break (strength was not scaled) and the movement of units during tests was not reported.

   Breakage of one unit in the armour layer may cause larger forces to occur on adjacent units as they re-adjust; moreover, the broken unit is often easily moved out of place

because of the loss of interlocking and the reduced weight of each piece. This allows units adjacent to the broken one to move more easily, resulting in impact with other units, causing large dynamic forces which may result in additional breakage of units. Breakage of one or two units may result in increased breakage in the armour layer which could eventually result in an unravelling of the armour layer causing rapid failure of the unprotected structure. Breakage of armour units has been observed at a number of large breakwaters including Sines, Portugal; San Ciprian and Bilbao, Spain; Humboldt Jetties, Crescent City, Cleveland and Kahului, U.S.A.; Baie Comeau and Riviere-au-Renard, Canada; Kochi and Kametoku, Japan, and Gaansbai, South Africa.

Many of the rubble-mound breakwaters existing today that have not been damaged, contain an armour layer designed to be stable without the assistance of interlocking between units. A number of these breakwaters contain concrete cubes. It is on these breakwaters, where the stability is not dependent upon interlocking, that little movement of the units occur, and where the structural consequences of interlocking do not have to be considered.

On some breakwaters, the armour layer consists of regularly placed blocks such as tribars, seabees, cobs, sheds, etc., that depend on friction achieved between the nested units. These structures are susceptible to settlement and movement of the armour layer which may result in individual unit breakage. Serious damage to these structures has been reported at Diablo Canyon and Nawiliwili in the U.S., for example.

7. Wave Overtopping

   Waves overtopped the structure and damaged the back and unprotected side of the structure.

8. Scour

   Wave action caused scour at the base of the breakwater allowing settlement of the armour stone.

9. Hydrostatic Pressure in the Rubble Mound

   Wave action caused large, internal and dynamic water pressures in the rubble mound. These pressures reduce the strength of the core material which may lead to instability of the mound.

10. Protection Below the Armour Layer

   The exposed surface of the rubble mound immediately below the armour layer was damaged, leading to settlement or collapse of the armour layer.

6 DESIGN PROCEDURES FOR THE EFFICIENT AND SAFE DESIGN OF AN ARMOUR LAYER

6.1 Introduction

A safe and cost-efficient breakwater may not result from use of the procedures that have been proposed and used in the past.

A successful design procedure will involve an experienced engineer undertaking the following activities:

i) The accurate definition of the environmental conditions for which the breakwater must be designed to withstand

ii) Careful and detailed study of the site and local materials and their properties and availability

iii) Review of possible construction procedures that will be suitable for the site and the available equipment that can be provided by the local community

iv) Development of trial designs using well designed and properly interpreted model studies of wave action on the breakwater

v) Review of the following aspects of the developed designs:

- optimum use of available materials

- ease of construction

- stability, overtopping, scour

- mode of failure if design conditions are exceeded and ultimate capacity of the breakwater

- cost

- acceptance by local community

If the procedures discussed in this section are followed, it is unlikely that the resulting structure will be similar to conventional concepts described in much of the technical literature.

The development of trial designs (item iv) for armour layers using quarried stones or concrete units are discussed in the following sections.

6.2 Breakwaters With Quarry Stone Armour

The design of a breakwater with an armour layer of quarried stone will follow the generalized procedure outline in Section 6.1. Some special considerations will be as follows:

- Properties of the Available Quarried Stones
It must be determined that the available stones are of sound, durable material, free of cracks and splits.

- Gradation of the Armour Layer
Significantly increased stability may be

achieved if a wide range of stone sizes are used in the armour layer. There is no reason for using stones of one size and obtaining them may involve inefficient use of the quarry yield.

- Thickness of the Armour Layer
If armour stones are available in large quantities, a significant increase in stability may be achieved by using an armour layer that is wider than the diameter of two stones.

- Enhanced Stability Produced by Wave Action
Designs should be considered where wave action early in the life of a structure causes increased stability by consolidating the armour layer, and by sorting the exposed stones into an interlocking layer.

- Placed Stone Technique
The use of placed stone techniques may offer advantages. The designer must ensure that an effective quality control procedure is established for the selection of the armour stones so that the armour layer is built without any flaws. Diver's inspection below the water may be required to ensure that the required placement technique is achieved. The design of the details at the base of the armour layer must not permit any settlement of the armour layer. At many locations, these may be unrealistic construction requirements. Construction of this concept is suited to relatively shallow water locations where there are long periods of calm.

It is considered that the full potential of armour stones in protecting a breakwater has not been realized. Armour layers consisting of quarried stones could be used at many locations where more expensive and problematic concrete units have been used in the past, if innovative designs are used in place of traditional concepts.

6.3 Breakwaters With Concrete Armour Units

The design of a breakwater with an armour layer of concrete armour units will also follow the generalized procedure outlined in Section 6.1.

However, for concrete armour units special consideration must be given to the optimum geometry of the unit and to its structural design. The development of the required armour unit will follow the generalized activities that are summarized in Figure 9.

Initially, a preliminary design is selected where the armour units are not displaced from the armour layer under the design wave conditions. This initial selection may be based on experience or the formulas previously discussed. The assumption is made at this point that the armour has sufficient strength to resist the applied loadings.

For the selected design, the loadings that occur on individual armour units are then determined. These include dynamic loadings resulting from wave action on the structure, from collision between adjacent units and from projectiles consisting of broken pieces of units, and static loadings that are imposed by adjacent units or as a result of differential settlement of the structure. A selected extreme loading will have an associated probability of occurrence. This is the most complex of the activities to be completed.

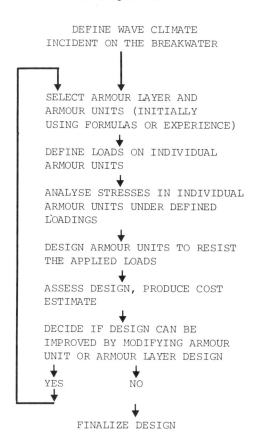

DEFINE WAVE CLIMATE
INCIDENT ON THE BREAKWATER

SELECT ARMOUR LAYER AND
ARMOUR UNITS (INITIALLY
USING FORMULAS OR EXPERIENCE)

DEFINE LOADS ON INDIVIDUAL
ARMOUR UNITS

ANALYSE STRESSES IN INDIVIDUAL
ARMOUR UNITS UNDER DEFINED
LOADINGS

DESIGN ARMOUR UNITS TO RESIST
THE APPLIED LOADS

ASSESS DESIGN, PRODUCE COST
ESTIMATE

DECIDE IF DESIGN CAN BE
IMPROVED BY MODIFYING ARMOUR
UNIT OR ARMOUR LAYER DESIGN

YES   NO

FINALIZE DESIGN

DEVELOPMENT OF A RANDOMLY PLACED CONCRETE UNIT

FIGURE 9

In the next stage of the design process, the armour unit is structurally designed to resist the accepted extreme loading conditions. This process involves numerical procedures for determining the stress distribution in the unit resulting from the defined loads. This is not a complex task; it involves developing a methodology suitable for the selected unit and expected loadings. A dolos unit discretized for analysis using finite element methods is shown in Figure 10.

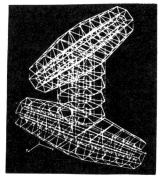

FIGURE 10

Conventional structural engineering procedures are used to design the armour unit to resist the defined stress distributions. This may include the use of reinforced concrete, fibre reinforced concrete, prestressed concrete or the use of an alternate material.

Once the unit is designed, an assessment is made as to whether the overall breakwater or the armour unit design could be improved. For example, if the armour layer is built on a flatter slope, a reduction in the magnitude of the loads applied to individual units may occur or the geometry of the armour unit may be modified to reduce the maximum value of stress in the unit. If changes to the design are made, then the process of defining the loads is repeated for the new design. Finally, an optimum and safe design results.

Regularly placed armour units require additional investigation. For this concept, the integrity and stability of the complete layer must be assessed. Special consideration must be given to the top, bottom and sides of the layer to ensure that failure will not initiate at these locations. Any loss of stability due to curvature of the armour layer must be determined. The consequence of breakage or removal of one or two units and the possible unravelling of the layer should be assessed. The analysis of loads on individual units will concentrate on the forces transmitted between the units giving special consideration to the situation where differential settlement of the underlayer occurs.

REFERENCES

1. Rennie J., "The Theory, Formation and Construction of British and Foreign Harbours", Weale, London, 1854.

2. Stevenson T., "The Design and Construction of Harbours", Black, Edinburgh, 1864.

3. Matthews W., "Harbours of Great Britain", Transaction of American Society of Civil Engineers, Vol. LIV, Part A, pages 159-180, 1905.

4. Rennie J., "An Historical, Practical and Theoretical Account of the Breakwater in Plymouth Sound", Weale, London, 1848.

5. Cunningham B., "A Treatise on the Principles and Practice of Harbour Engineering", Griffen, London 1928.

6. Iribarren C., "Una Formula par el Calculo de Digues de Escollera", Revista de Obras Publicas, 1938.

7. Hudson R. Y., "Design of Quarry-Stone Cover Layers for Rubble Mound Breakwaters", US Army Corps of Engineers, W.E.S., Research Report 2-2, 1958.

8. Anon, "Final Report of the International Commission for the Study of Waves", Permanent International Association of Navigation Congresses Bulletin No. 25, Vol. III, 1976.

9. Kingman D.C., "Harbours on Lakes Erie and Ontario", Transaction of American Society of Civil Engineers, Vol. LIV, Part A, pages 237-262, 1905.

10. Debok D.H. and Sollitt C.K., "Large-Scale Model Study of Placed Stone Break-waters", Oregon State University Report, 1973.

11. Markle D.G. and Davidson D.D., "Placed-Stone Stability Tests, Tillamook, Oregon", US Army Engineer Waterways Experiment Station, Technical Report HL-79-16, September 1979.

12. Hall K.R., Rauw C.I., Baird W.F., "Develop-ment of Wave Protection for an Offshore Runway Extension at Unalaska Airport, Alaska", American Society of Civil Engineering Specialty Conference, Coastal Structures 83, March 1983.

13. Shima S., "Concrete Blocks at Osaka Harbour Works", Transactions of American Society of Civil Engineers, Vol. LIV, Part A, pages 221-236, 1905.

14. Scott M., "Description of a Breakwater at the Por of Blyth;and of Improvements in Breakwaters Applicable to Harbours of Refuge", Proceedings of the In-stitute of Civil Engineers, Vol. 18, 1858-1859, p72.

15. Scott M., "On Breakwaters, Part II", Pro-ceedings of the Institute of Civil Engineers, Vol. 19, 1860-1861.

16. Jackson R.A., "Design of Quadripod Cover Layers for Rubblemound Breakwaters", Miscellaneous Report 2-372, US Army Waterways Experiment Station, 1960.

17. Hudson R.Y., "Protective Cover Layers for Rubble Mound Breakwaters, Studies Completed Through March 1957", Misc. Paper 2-276, US Army Waterways Experi-ment Station, July 1958.

18. Nagai, "Study of Hollow Tetrahedrons for Absorbing Wave Energy", Proceedings of 7th Conference on Coastal Engineer-ing, Tokyo, 1960.

19. Singh K., "Stabit, A New Armour Unit", Pro-ceedings of the 11th Conference on Coastal Engineering, London 1968.

20. Fisheries Development Corporation of South Africa, "Breakwater Armour Units - Model Tests", Research Dept., May, 1974.

21. Paape and Walter, A.W., "Akmon Armour Unit for Cover Layers of Rubble Mound Breakwaters", Proceedings of 8th Conference of Coastal Engineering, Mexico, 1962.

22. Palmer R.Q., "Breakwaters in the Hawaiian Islands", ASCE, Journal of Waterways and Ports, Vol. II, No. 13, Dec. 1952.

23. Brown C.T., "Seabees - A Third Generation Armour Unit", 7th Harbour Congress, Vol. 1, Section 1 and 2, Antwerp, May 1978.

24. Coode & Partners, "Artificial Armouring of Marine Structures", Dock and Harbour Authority, November 1970.

25. Shepard Hill Ltd., "The SHED Armour Unit", Internal Report, Shepard Hill Ltd., Civil Engineering Contractors, June 1982.

26. Ahrens, J.P., Large Wave Tank Tests of Riprap Stability", Tech. Mem. No. 51, Coastal Engineering Research Center, Fort Belvoir, Virginia, May 1975.

27. Hudson R.Y., "Design of Quarry Stone Cover Layers for Rubblemound Breakwaters", Research Report 2-2, US Army Water-ways Experiment Station, July, 1958.

28. Dai Y.B. and Kamel A.M., "Scale Effects Tests for Rubblemound Breakwaters", Research Report H69-2, US Army Water-ways Experiment Station, Dec., 1969.

29. Decavalho J., "Praia Da Vitoria Harbour (Axores) - Damages Due to the Storm of December 26, 27, 1962", Proceed-ings of the 9th Conference on Coast-al Engineering, held in Washingon, published by ASCE, New York, N.Y., 1964.

30. Magoon O.T., Sloan R.L., and Foote G.L., "Damages to Coastal Structures", Proceedings of the 14th Coastal Engi-neering Conference, held in Copenhagen published by ASCE, New York, N.Y., 1974.

31. Takeyama H. and Nakayama T., "Disasters of Breakwaters by Wave Action (2)", Port and Harbour Research Institute, Tokyo, Japan, 1975.

32. Magoon O.T. and Baird W.F., "Breakage of Breakwater Armour Units", Proceedings of the Symposium on Design of Rubble Mound Breakwaters, British Hovercraft Corporation, Isle of Wight, 1977.

33. Foster D.N., "Model Simulation of Damage to Rosslyn Bay Breakwater During Cyclone David", Proceedings of the 6th Australian Hydraulics and Fluid Mechanics Conference, Adelaide, 1977.

34. Standish-White D.W. and Zwamborn J.A., "Problems of Design and Construction of an Offshore Seawater Intake", Proceedings of the 16th Coastal Engineering Conference, held in Hamburg, published by ASCE, New York, N.Y., 1978.

35. Bruun P., "Common Reasons for Damage or Breakdown of Mound Breakwaters", Coastal Engineering, Vol. 2, 1979.

36. Zwamborn J.A., "Analysis of Causes of Damage to Sines Breakwater", Proceedings of the Special Conference on the Design, Construction and Maintenance, and Performance of Port and Coastal Structures, Vol. 1, ASCE, New York, N.Y., 1979.

37. Zwamborn J.A., Bosman D.E. and Moes J., "Dolosse - Past, Present, and Future", Proceedings of the 17th International Conference on Coastal Engineering, held in Sydney, published by ASCE, New York, N.Y., 1980.

38. Losada M., "Spanish Experience with Concrete Armour Units", Proceedings of ASCE Convention, ASCE, New York, N.Y., 1981.

39. Pita C., "Prototype Experience With Concrete Armour in Portugal", Proceedings of ASCE Convention, ASCE, New York, N.Y., 1981.

40. Sullivan S., "Experience with Concrete Armour in the Hawaiian Islands", Proceedings of ASCE Convention, ASCE, New York, N.Y., 1981.

41. Clark D., "U.S. Great Lakes Practice with Concrete Armour Units", Proceedings of ASCE Convention, ASCE, New York, N.Y., 1981

42. Baird W.F., Hall K.R. and Magoon O.T., "The Design of Concrete Armour Units", Proceedings of ASCE Convention, ASCE, New York, N.Y., 1981.

Discussion on Papers 7–9

P. ACKERS and J. D. PITT (*Introduction to Paper 7*)
The usual methods of designing wave protection
involve determining the wave climate for a
selected return period of storm, selecting
armour size in accordance with a formula relating size to wave height; knowing the size, the
layer thickness is assessed and then various
ancillary calculations are made for the grading
and thickness of underlayers, upper limit to
accommodate run up etc.

Most design relationships show armour size
(linear dimension) to be proportional to wave
height, and layer thickness to be a multiple
of armour size. So tolerances on the design
calculations give rise to at least an equal
tolerance on costs. A 50% overestimate of armour
size leads to at least a 50% overspend; but by
the same token a 20% saving on costs by over-
optimistic design would be exceedingly dangerous.

Hence there is a need for accurate design
information. In this context, uncertainty in
scale effects when converting results from
laboratory tests to design information can not
be glossed over, and this factor provided the
stimulus for the field research described.

Any field project relying on natural events
is a risky business, but we were favoured by
having some major storms in a reasonably short
period covering the full range of damaging
events for which the test panels were designed.
Apart from the direct benefit of comparing
damage with hindcasts from observed waves, using
the laboratory-based damage functions, the field
exercise helped concentrate attention on some
practical problems of slope protection.

Layer thickness. Thickness is usually specified as a multiple of size, but we found the
stone sizes generally to be larger than specified so that layer thickness ratios were propor-
tionately reduced. It is also difficult to
assess layer thickness and the answer is dependent on the surveying method.

Segregation. In coastal works, multiple
handling of armour stone is inevitable, from
quarry face to its final position. This leads
to segregation so that parts of the layer as
placed may be undersized, and parts oversized.
Subjective adjustment after placement is of
limited benefit.

Grading. The practice in laboratories of
taking a large representative sample and sieving
it can not be duplicated in the field. Any
sampling and grading is difficult, time-consuming
and involves mechanical handling and weighing.
So it is not surprising that the gradings of
armour stone actually used for slope protection
may not be known precisely in many schemes and
that control from the quarry may not be as
close as one would wish.

When one adds these practical difficulties to
those of assessing the design wave and uncertainty in the function of stone size to wave
height, the occasional weakness or even failure
is understandable - although our objective must
be to minimize those risks.

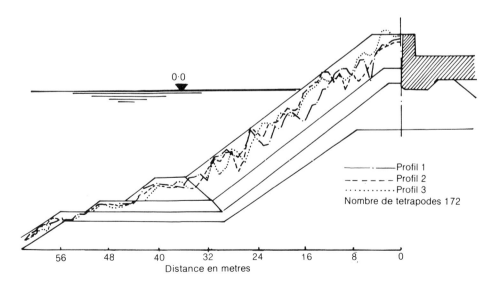

0·0

Profil 1
Profil 2
Profil 3
Nombre de tetrapodes 172

56   48   40   32   24   16   8   0
Distance en metres

Fig. 1. Surveys made in a model of the breakwater in Port d'Arzew El Djedid

Breakwaters—design and construction. Thomas Telford Ltd, London, 1984

121

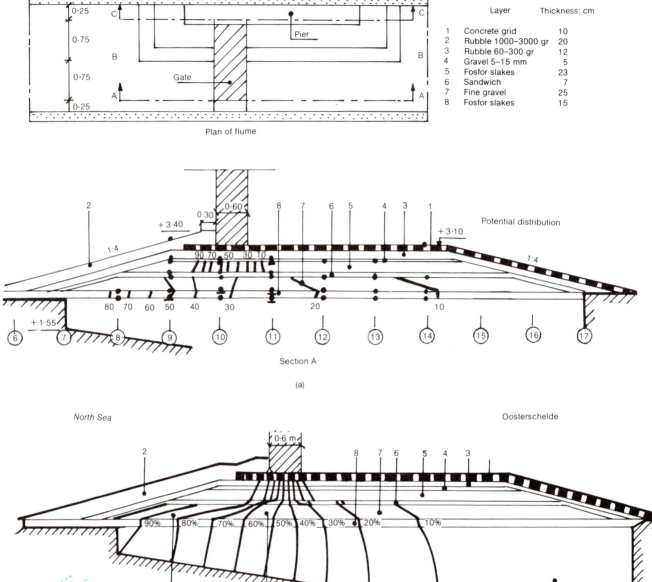

Layer		Thickness: cm
1	Concrete grid	10
2	Rubble 1000–3000 gr	20
3	Rubble 60–300 gr	12
4	Gravel 5–15 mm	5
5	Fosfor slakes	23
6	Sandwich	7
7	Fine gravel	25
8	Fosfor slakes	15

Plan of flume

Section A

(a)

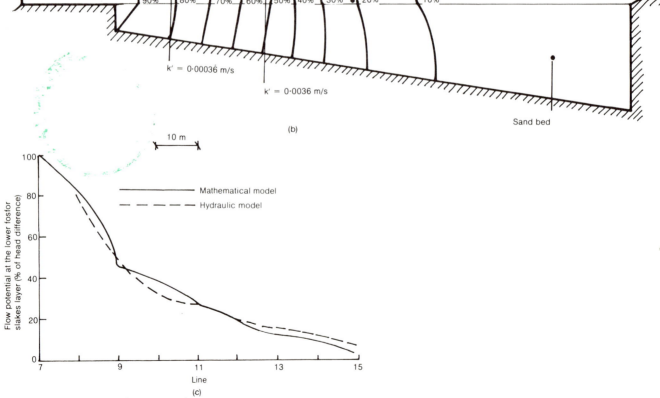

(b)

(c)

Fig. 2. Turbulent flow verification: (a) results of the hydraulic model; (b) potential distribution computed for the configuration and materials of the hydraulic model; (c) comparison between the mathematical and the hydraulic model

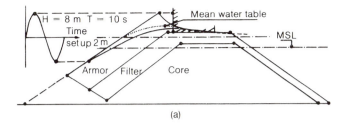

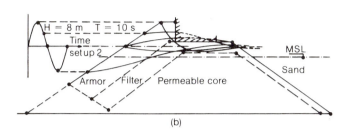

Fig. 3. Internal set-up on a breakwater: (a) watertable for storm wave conditions; (b) water-table fluctuation for storm wave conditions and a permeable core

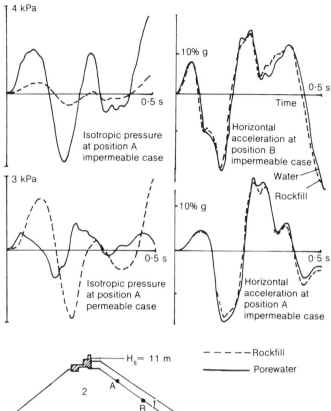

Isotropic pressure at position A impermeable case

Horizontal acceleration at position B impermeable case

Water

Rockfill

Isotropic pressure at position A permeable case

Horizontal acceleration at position A impermeable case

Fig. 4. (right). Effect of permeability on the dynamic response

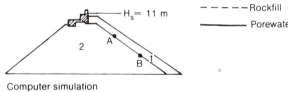

H_s = 11 m

- - - - Rockfill
——— Porewater

Computer simulation

Filter 1 K = 0·33   K = 0·033 m/s
Core 2 K = 0·033   K = 0·0033 m/s
        Permeable Impermeable

Dynamic response due to impact wave

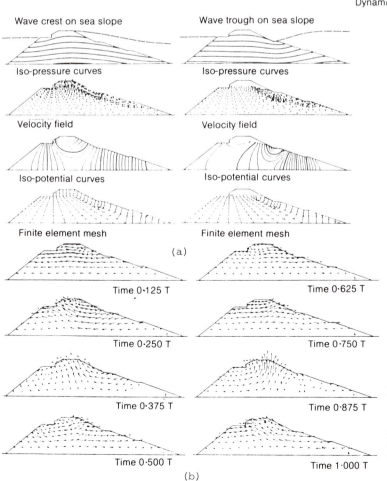

Wave crest on sea slope

Iso-pressure curves

Velocity field

Iso-potential curves

Finite element mesh

(a)

Time 0·125 T

Time 0·250 T

Time 0·375 T

Time 0·500 T

(b)

Wave trough on sea slope

Iso-pressure curves

Velocity field

Iso-potential curves

Finite element mesh

Time 0·625 T

Time 0·750 T

Time 0·875 T

Time 1·000 T

Fig. 5. Total behaviour of a breakwater simulated by numerical models: (a) transient turbulent porous simulation (SEEP and HADEER codes); (b) deformation modes due to a wave impact, main period about 0.5s (SATURN code)

O. J. JENSEN (*Danish Hydraulic Institute*)
Surveying of breakwater profiles is very important both during and after construction of new
structures. For damaged breakwaters this is also
very important.

In order to visualize the difficulties in
planning and interpretation of such surveys,
Fig. 1 shows the results of three repeated
surveys of a model of the Port d'Arzew El Djedid
breakwater built according to the project
drawings. In the model the survey was made in
the same way as previously done in practice, by
using a 30 cm device suspended by a string. The
soundings showed the depth at which this device
touched the tetrapods (20 m^3, height 4.15 m)
or quarry stones in the slope.

The measured tetrapod 'surface' is about 2 m
below the theoretical one and the 'surface' is
highly irregular.

This example demonstrates that the method of
sounding of rubble-mound breakwaters should be
tuned to the 'roughness' of the surface to be
measured. If possible similar measurements
should be made in a scale model to be used as
reference for the prototype conditions, but
under more controlled conditions.

H. LIGTERINGEN (*Delft Hydraulics Laboratory*)
What was the underlayer porosity for the regular
slope protection on both sides of the test
panels?

Was an attempt made to reproduce the stability
of this regular slope protection (including the
porous underlayer) in the model tests after the
full-scale tests? This could give an answer to
the question of how far the underlayers contribute to - or are the sole reason for - scale
effects.

What is the global estimate of the cost of
this long year programme?

F. B. J. BARENDS (*Delft Soil Mechanics
Laboratory*)
Geotechnics in breakwater design: In the last
few decades harbours have been constructed in
deeper waters and the height of breakwaters has
increased. The volume of these structures have
become large and consequently the soil-rock
mechanical behaviour plays a more pronounced role
in the overall stability. The design approach
which conventionally considers the hydraulic
stability of the armour and underlaying filter
layers needs to be extended accordingly.

Recent failures of several large breakwaters
(ref. 1) prompted studies on the local and
total mechanical stability of the structure in
an extreme wave-loading environment, and studies
concerning the constitutive behaviour of rockfill and the applicability of a continuum
approach.

Three fundamental aspects can be distinguished:

(a) low-frequency loading, characterizing
    internal turbulent porous flow generated by
    waves (seepage)

(b) shock loading, typifying the dynamic response
    due to wave impacts

(c) rock-mechanical characteristics at collapse
    (slip surface analysis)

The expertise gained from studies on the
coarse granular sill in the projected storm
surge barrier in the Eastern Scheldt Estuary in
the Netherlands (ref. 2) resulted in a well
calibrated numerical model to simulate turbulent
porous flow, the SEEP code (Fig. 2). Recently,
a hybrid method, the HADEER code, was completed
and calibrated to several large-scale tests
(ref. 3). The predicted effect due to geometric
non-linearity, kinetic energy dissipation, and air
entrainment, which causes a significant internal
set up, was confirmed by physical model tests
(Fig. 3). This effect actually embodies the
influence of low-frequency loading on the
mechanical stability.

In a similar fashion the dynamic response due
to a wave impact at the crown structure has
been studied applying a sophisticated numerical
model, the SATURN code, suited to simulate the
true dynamic two phase (porewater and rockfill)
medium-structure interaction (ref. 4). To
verify the outcome of these calculations physical
model tests have been carried out, equipped with
special devices to detect generated accelerations
and dynamic porewater pressures. The measurements underscored the computationally discovered
phenomena: serious dynamic porewater pressures
and accelerations, which essentially affect the
mechanical stability of the breakwater under
wave impact conditions (ref. 5).

Wave impact energy is conveyed through the
structure with little or no damping at all,
when the breakwater is highly saturated. The
reaction of a two-phase medium (water and rockfill) strongly depends on the local permeability
(Fig. 4). Measurement of the dynamic porewater
pressures reveals only a part of the response.
The accelerations generated in both media are
almost equal, hence deformations of the core
material are likely to occur. In fact the
entire breakwater responds to the impact (Fig. 5)
whereas locally - particularly at the top -
irreversible deformations occur: a cyclic dege-
neration of the structure (ref. 5).

The deformation at failure state is highly
dependent on the internal friction mobilized
in the grain skeleton formed by the rockfill
body of the breakwater. The internal friction
behaviour of rockfill material is, however.
essentially different compared with that of soils
like sand and clay. For rockfill the rock
stiffness, in situ stress level and roughness of
the grain surface show a pronounced effect on the
internal friction. A conventional slip-surface
analysis model, the STAGROM code, has been
adapted with a special facility to account for
this particular phenomenon (ref. 5). Moreover,
dynamic porewater pressures generated by waves
and accelerations caused by wave impacts can be
automatically accounted for in the stability
analysis. In this manner a sound and more
complete insight into the geotechnical stability
of rubble-mound breakwaters has been obtained.

If the subsoil on which the breakwater is
situated contains a sand layer, one has to
consider the fact that it responds to wave
pressures. Some sands can lose consistency and
liquefy. In such a case the breakwater toe may
gradually collapse allowing the armour layer to
slip down. The unprotected core at the top is
open to direct wave attack, and the afore-
mentioned failure mechanisms will occur even

more rapidly. Special laboratory tests on carefully obtained in situ samples are to be performed to evaluate the liquefaction potential in order to obtain realistic soil resistance parameters. In situ density measurements are to be performed to assess the toe stability, or one should provide a design, that eliminates this risk (geotextile). The overall geotechnical stability analysis is not complete without a good investigation of the toe stability.

An essential aspect to be highlighted concerns the degradation of the rubble-mound mechanical properties: internal erosion of the core and the filter layers. This feature is as important as the breakage of armour units, which attains a great deal of attention. Some quarry stone may disintegrate gradually but drastically. The composition and functioning of a breakwater may change accordingly. One believes that a first natural settlement of the structure during the initial storms will densify the system to make it more reliable to withstand future design storms. Is this generally true? It depends on the rockfill quality. An investigation of the long-term constitutive behaviour of rockfill, and also weak rock available in the nearby neighbourhood (including its natural degradation), will give an answer. A surprise after 5 or 10 years can be foreseen.

The design of breakwaters should include the geotechnical aspects involved in the behaviour of the structure as a whole under relevant dynamic conditions. The methods and means to evaluate these aspects exist. An optimum design procedure with regard to geotechnics is available.

P. HUNTER (*Sir Alexander Gibb & Partners*)
Different results have been obtained from using two equally conventional methods of designing the underlayer beneath rock armour.

(a) Weight method (ref. 6)
Armour weight = w

$$\text{Underlayer} = \frac{w}{10} \pm 30\%$$

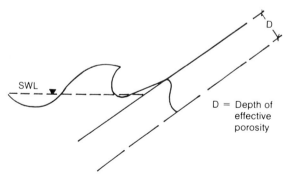

Fig. 6. Porosity/dissipation relationship

(b) Method derived from Terzaghi's Filter rules, originally for steady flow but modified for oscillating flows beneath rock armour:

Modified filter rules method (ref. 7)

$$D_{85F} \geqslant \frac{D_{15A}}{5}$$

$$D_{50F} \geqslant \frac{D_{50A}}{7}$$

$$D_{15F} \geqslant \frac{D_{15A}}{7}$$

For an armour layer having a mean weight of 5 t, the underlayer becomes 330-650 kg by the weight method, and 15-70 kg by the modified filter rules method.

If both approaches are valid, it is possible to allow a wider range of sizes in underlayers than is commonly used. This would allow economies in quarry operations and sometimes in the number of intermediate layers.

However, there are two points for caution: Firstly, reduced friction between armour and filter; although I suspect this failure mechanism is only relevant on very steep slopes armoured with a thin layer of rock.

Secondly, the need to project the core before the armour has been placed; although if the use of a coarser underlayer leads to a need for a further (finer) sublayer, then this benefit is largely lost.

It appears that the accuracy with which grading limits of underlayers are specified is not very critical. It is possible that failures due to internal erosion are caused by layers being too thin in comparison to realistic tolerances which can be achieved underwater. Sometimes layers of 0.5 m thickness or less are specified. This could lead to occasional holes occurring in the underlayer.

I would suggest that no underwater layers should be less than 60 mm thick, and in increasing depths or rougher conditions one metre is a more appropriate minimum.

E. H. HARLOW (*Consultant, Texas*)
Mr Hedges referred to dynamic pressures of breaking waves. Groeneveld and Jensen (ref. 8) gave measurements of these pressures. One can calculate their effect on pore pressures in the core!
Philosophically, our problem with breakwaters in deeper water is that rock cannot be obtained large enough to be stable in attack by big waves, and concrete armour is generally not strong or durable enough.

Table 1. Solids Ratio

	3D	2D
Natural Armour Stone	60	60
Diodes Mk I	41	66
Diodes Mk II	34	62
Reefs	19	50

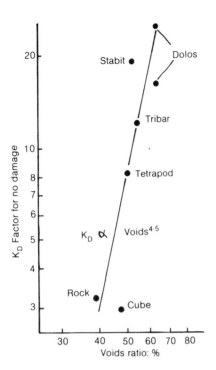

Fig. 7. Relationship between block stability and voids ratio

Does armour really 'protect' the rubble mound? It does cause turbulence which limits run up and overtopping. If we had real protection in the form of an impervious membrane over the seaward slope, we would avoid the pore pressure effects - but we would have very high structures to avoid overtopping.

The conditions stated by Mr Hedges for core material could well be met by low-strength rolled concrete, as was employed in a dam described during previous discussion. In fact, a breakwater is partly a dam, a two-sided revetment, with one side facing turbulence, and the other laminar conditions. In dams, the interiors are built under controlled conditions, but in breakwaters, very little. In one study of offshore islands, a few years ago, the most economical sea defence system was found to be a flat, sand slope. If one could assure that the sand moved only up and down the slope, not laterally, one could be content with such a design.

One fact seems clear - if the exterior of a breakwater is rough, it must have a heart of gold.

P. C. BARBER (Metropolitan Borough of Wirral)
The importance of the control of porosity in the core and underlayers of breakwater construction has been identified and I should like to comment upon the importance of porosity with regard to the primary armouring of a structure.

The performance of a primary armour may be summarized under the properties of: stability, reflection/transmission, run up and structural integrity. All these properties depend on the porosity of the construction.

Work on porosity has been carried out for the works of coastline control on the Wirral over the last six years in association with Liverpool University and now Imperial College of Science and Technology. The work has shown

that wave-form dissipation, affecting reflection and transmission, is dependent upon three-dimensional porosity over the effective depth of active construction with the specific surface used for porosity development and that flow dissipation, affecting run up, is dependent upon two-dimensional porosity over the effective depth of active construction with the specific surface used for porosity development (see Fig. 6).

These dependences were established from model tests carried out in the random wave flume at Wallingford and have resulted in the development of the DIODE wave absorption unit for slope armouring and the REEF unit for crest-armouring of submerged breakwaters.

The diode mark I unit, laid at a 1 in 2 slope, exhibited a reflection coefficient of 0.20 and a run-up coefficient of 0.6. The unit weight was 2 t with a depth of 1.10 m and typical plan dimension of 2.0 m and was stable up to the maximum significant wave heights of 5.0 m available for the model scale. These units have been manufactured and installed and during February 1983 they withstood a once in fifty year event successfully. A mark II version of the unit is at present under design for manufacture and installation next year. This unit will have a greater three-dimensional porosity to improve the reflection performance of the construction and the run-up performance will be maintained by retaining the two-dimensional porosity.

A surface-piercing slope, which is successful in wave dissipation, attracts wave set-up which results in larger waves reaching the structure, especially if the set-up dissipates by current parallel to incident wave crests. To overcome this effect it was decided to examine the potential of submerged breakwaters for the works of coastline control. Previous work with natural armour stone structures was not encouraging in performance, but when experiments were undertaken with different porosities of crest armour in two- and three-dimensional format the performance improved. The reef unit weighs about 6 t and is of 2.5 m typical dimension. When laid to pattern for model testing at the Hydraulics Research Station Ltd, it reduced transmitted energy by 60%, with a 2.0 m clear water depth above the crest in a general water depth of 8.0 m for an incident sea state of 3.0 m significant wave height; the reflection coefficient was less than 0.10. For this work the flow dissipation porosity is increased for gravity effects and the wave dissipation porosity is considerably increased. The various solids ratios are summarized in Table 1 for the units described above and compared with values for natural armour stone. Reef units are at present under manufacture and are due for installation in the near future.

C. D. HALL (Netlon Ltd, seconded to Manchester University)
In paragraphs 29 and 30 of Paper 8 it is stated that geotextiles have been used principally for their filtering and separating qualities. Would Mr Hedges comment on applications which capitalize on the strength of geotextiles, for example by introducing reinforcing layers within or under the breakwater core material.

H. F. BURCHARTH (*Aalborg University, Denmark*)
The crushing characteristics of the interblock contact points are of great importance for the assessment of internal geotechnical stability in terms of slip-circle stability, settlements etc. Attention should be paid, in a realistic and practical way, to the problem of determining the 'contact point crushing characteristics' of core and filter layer stones.

J. F. MAQUET (*Sofremer, Paris*)
Curves were presented by the Author showing the variation of friction coefficient between armour layer and underlayer according to the type of underlayer.

The underlayer should also be adapted according to the type of armour unit. Underlayer specifications should differ if the armour unit is a cube, a tetrapod or a dolos.

Thus, another question to be answered before design of underlayers should be 'What type of armour will be placed over the underlayer?'.

A. F. WHILLOCK (*formerly Hydraulics Research Station Ltd*)
Although a great deal of attention has been paid to the weight, size and shape of armour units it seems to have been overlooked that the work of dealing with incident wave energy takes place within the spaces between them.

Figure 7 is derived from ref. 9 and is similar to one published in ref. 10. It suggests that block stability increases as at least the fourth power of the voids ratio with remarkable consistency from rock to dolos. It is not possible to be sure if this is a cause or an effect; I would like to see tests conducted to find out. There are exceptions; the lower K_D value for the dolos block is from tests on a slope with a lower permeability than usually assumed. This shows that to obtain full benefit from enhanced voids in the cover layer there should be a commensurate improvement further in. The superior stability of the stabit on this basis must be due to some of the wave energy being dissipated within the frame of the block and the uplift correspondingly reduced.

It now seems clear that there is a limit to the size and complication of block design on structural considerations. A block with interlocking projections can be so large that it cannot support its own weight, let alone participate in the stresses of a random assembly under wave attack. Even the equivalent size of a compact form such as a cube, supposing it could be cast and handled, might prove equally fragile. Perhaps the availability of high stability blocks, judged by their hydraulic performance alone, has encouraged the construction of breakwaters in situations that could not be considered for simpler blocks.

I suggest, however, that there should be a return to a more compact, structurally sound type of block but with perforations so that a proportion of wave forces can be balanced internally without contributing to the uplift. Cubes are an obvious candidate but these have a habit of settling into regular formations unless precautions are taken. A cubical bipod or tripod would ensure that a random assembly would be maintained without difficulty.

From other considerations it may be that some of the difficulties are man-made. There is no need to meet trouble head on! The whole topic of breakwater and seawall design has been dominated by a two-dimensional approach which allows for analytical studies and economical testing at large scales.

Although there is a weakness of blocks with projecting limbs to an oblique wave approach (ref. 11), this is not found with compact forms. Recent tests conducted with oblique approaches to sea walls (ref. 12) indicate that high normal forces are halved at 15°. It is likely that forces on and in a random block breakwater can be reduced by aligning it so that a major wave front will not strike uniformly along its length. The wave energy will thus be dealt with over a longer time, a larger area and at a lower level.

For economical reasons, a breakwater is often aligned along a contour which itself has been set normal to the dominant wave approach. This was the case with Stevenson's breakwater at Wick (ref. 13), which was merely waiting for a wave of the right height to come along. Other alignments would be more expensive but the cost of safety has to be met somewhere.

To combine the foregoing ideas, a sound design for an exposed site would be a slope not steeper than 1 in 2 of compact blocks, perforated for improved wave energy absorption, and aligned so that a major wave attack will not occur at less than, say, 20° to the normal. In deep water a large caisson roundhead would be more practicable and certain than a broad rubble slope and, if perforations do not fulfill their promises, a caisson trunk will be required also. At the shore end a flat rubble slope will deal with residual and deflected waves, where maintenance would not be difficult.

Such a structure might well be rejected by man as too costly but not by nature as inadequate for its purpose.

W. A. PRICE (*Hydraulics Research Station Ltd*)
In Paper 8 stability was listed as one of the most important aims in design and I am sure that in most people's minds this means that the rock or armour units should not move. But there are features in nature like breakwaters (e.g. Chesil Beach) which are dynamically stable. In storms a great deal of movement would be taking place. Hence when the quarry cannot supply the material to fit Mr Baird's criteria, then all is not lost. It might still be adequate to form a good rock beach. Such a structure will demand maintenance but perhaps not so much as would be expected.

So returning to Paper 8, I would alter stability to 'static or dynamic stability'.

E. G. WISE (*Edgar G. Wise & Associates*)
The types of armouring, so far considered and discussed have shared one common concept; namely that the units comprising the armoured surface, whether of natural rocks or prefabricated forms, are all separate and discrete units whose individual effectiveness is a function of a considerable localized mass and a variable and indeterminate degree of interlock.

Mention has also been made of difficulties

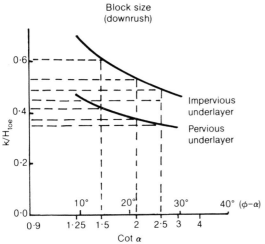

Fig. 8. The dimensionless ratio k/H_{toe} as a function of cot α. (The diagram is valid for ratios of densities γ_s/γ_f = 2.65/1.00; 2.70/1.02 or 2.75/1.035 and the internal friction angle ϕ = 48°)

and uncertainties in the underwater placing of the separate armouring units, and attention drawn to the merits of regularly placed cubical units, probably in single layers.

In recent times I have been associated with a number of anti-erosion and coastal protection projects, whose wave attack surfaces have been armoured by a single layer of cubical concrete blocks, integrated into a continuous structure by high tensile parallel woven polypropylene tendons passing through each block in both longitudinal and lateral directions. The system owes its attraction to recent developments in the manufacture of inert and durable tendons for which manufacturers are prepared to provide guaranteed minimum breaking loads.

I suggest that in the future the two-dimensional interconnection of single layer armour, and possibly also a three-dimensional interconnection of multiple layers may provide real economies in effort and time, with improved laying techniques; and may perhaps introduce some modification of the damage coefficient in the Hudson formula.

P. A. HEDAR (*P. A. Hedar AB, Gothenburg*)
Waves are going up and down when attacking a rockfill breakwater and are killed on the windward slope. There is a well-founded reason to divide the problem into two parts, one for the uprushing wave phase and one for the downrushing wave phase. In doing so the following stability formulae for the armour layer are obtained by theoretical derivation and small-scale tests.

$$Q = \frac{\P}{6} \gamma_s k^3 \text{ (the blocks equal balls, } k \text{ is a fictive diameter of the block)}$$

Table 2. Various features and effects on the durability of concrete with time, in a marine situation

FEATURE	POSSIBLE EFFECTS	DURATION - LOG TIME					TREATMENT
		1 week	1 month	1 year	10 years	50 years	
Plastic shrinkage. (inc. plastic settlement)	Cracking, localised loss of bond and / or cover to reinforcement.	P/C					Improve mix design. Start efficient curing quickly.
Bleeding. Honeycombing.	Loss of actual cover to reinforcement and damage to surface.	P					Improve mix design and workmanship. Improve aggregates.
Initial thermal contraction.	Cracking in depth, localised loss of cover to reinforcement.		P				Reduce peak temperatures and temperature differentials. Allow for movement.
Drying shrinkage.	Cracking in depth, localised loss of cover to reinforcement.			P			Improve mix, especially lower w/c ratio.
Unsound aggregates.	Surface disintegration and general loss of cover to reinforcement. Especially intertidal and splash zone.			P	C		Do not use. Test for first.
Erodable aggregates, low density concretes Carbonate beach sands.	Differential surface disintegration and general loss of cover to reinforcement Especially intertidal and splash zone.				P/C		Avoid poor aggregates, design for dense concrete.
Sulphate attack- external.	Possible expansion and cracking surface disintegration and loss of cover to reinforcement.			C			Design and achieve dense concrete
Physical salt weathering.	Surface disintegration and localised loss of effective cover to reinforcement. Especially splash zone and above.			P			Design and achieve dense concrete, use sound aggregate.
Excess internal chloride.	Accelerated reinforcement corrosion, cracking and spalling.				C ·		Use clean aggregates and potable mix water
Reactive aggregate.	Pop-outs, expansion and cracking, loss of cover to reinforcement.				C		Do not use. Test for first.
Loss of effective or actual cover to reinforcement.	Reinforcement corrosion, cracking and spalling.			C/P			Improve concrete as above, use dense concrete with deep cover to reinforcement.

C = Chemical effects

P = Physical effects

UPRUSH

Pervious underlayer

$$k = \frac{0.30 \; (d_b + 0.7H_b) \; (\tan\phi + 2)}{\left(\dfrac{\gamma_s}{\gamma_f} - 1\right) \left(\log_{10} \dfrac{14.83 H_b}{k}\right)^2 \cos\alpha \; (\tan\phi + \tan\alpha)}$$

Impervious underlayer

$$k = \frac{0.35 \; (d_b + 0.7H_b) \; (\tan\phi + 2)}{\left(\dfrac{\gamma_s}{\gamma_f} - 1\right) \left(\log_{10} \dfrac{14.83 H_b}{k}\right)^2 \cos\alpha \; (\tan\phi + \tan\alpha)}$$

DOWNRUSH

Pervious underlayer

$$k = \frac{2.32 \; H_{toe} \; (\tan\phi + 2)}{\left(\dfrac{\gamma_s}{\gamma_f} - 1\right) \left(e^{4\tan\alpha} + 13.7\right) \cos\alpha \; (\tan\phi - \tan\alpha)}$$

Impervious underlayer

$$k = \frac{3.70 \; H_{toe} \; (\tan\phi + 2)}{\left(\dfrac{\gamma_s}{\gamma_f} - 1\right) \left(e^{4\tan\alpha} + 16.5\right) \cos\alpha \; (\tan\phi - \tan\alpha)}$$

where

Q = weight of an individual block
k = a fictive characteristic of the linear dimension of blocks
d_b = water depth at breaking point
H_b = design breaking wave height (uprush)
H_{toe} = design wave height at the location of the proposed structure (downrush)
γ_s = density of rock
γ_f = density of fluid
ϕ = internal friction angle (degree of interlocking)
α = angle of armour slope
e = Napier's number

The formulae consist of a density part, a hydrodynamic part and an interlocking part; they illustrate the importance of the pervious underlayer and of the degree of interlocking.

If the original interlocking is increased in an artificial way, the interlocking will be reduced during the lifetime of the breakwater

by vibrations, wear and tear caused by waves and by differential settlements and penetration into sea bottom.

For short periodical waves and/or steep slopes the downrush formulae give the necessary block weight. For long periodical waves and/or flat slopes the uprush formulae will give the necessary block weight.

If the slope angle α approaches the internal friction angle ϕ, no strength is left for wave forces. This implies that the required block weight tends to infinity as can be seen from the formulae.

In Fig. 8, concerning downrush, the non-dimensional ratio k/h_{toe} is plotted against the difference between the internal friction angle and the slope angle $(\phi - \alpha)$.

Choice of the design wave height
Uprush: The design wave height is the breaking wave height H_b. The maximum breaking wave that can attack the breakwater is one created a half wave length in front of the toe.
Downrush: The design wave height is measured at the location of the proposed structure.
Significant wave height: H_s or H_o: It is open to question whether, for the design of a breakwater, the maximum wave height must be chosen, or some other wave height, e.g. H_{10} the average height of the highest one-tenth, or H_s the significant wave height must be chosen. This depends on the margin of safety demanded. The stability criterion is generally not a sharp limit and sometimes the movement of some stones may be tolerated, especially if repair is easy and cheap. If the maximum wave height at a certain frequency is chosen the breakwater will always be stable provided no abnormal wave height occurs. If an average wave height is chosen, some waves higher than the design wave height will occur. These higher waves must cause movements here and there.

Block weight category corresponding to the design wave height from H_s to H_{10} has up to now been used very often.

P. G. FOOKES (Consultant)
I would like to make two points concerning real full-scale situations. Firstly, in model studies of breakwaters with stone of a certain size and weight, which may or may not be interlocked, the only rounding which takes place is by stone movement rolling against each other. In a real breakwater rounding of armour stone by wave attrition takes place with time without the requirement of stones to roll against each other. A time may, therefore, be reached (after some years) where the stones will roll if they lose sufficient weight and loss of interlock (provided they had it in

Table 3. Variations in perceived thickness ratio

Panel	t:mm	(mm)	D_{50}(mm)	t/D_{50}	$(t+)/D_{50}$*	$(t+2)/D_{50}$**
1	440	75	230	1.9	2.24	2.57
2	480	120	400	1.2	1.50	1.80
3	570	137	500	1.1	1.41	1.69
4	760	143	560	1.3	1.61	1.87

*Considering only highest 40% of levels
**Considering only highest 6% of levels

the first place).

My second point concerns the deterioration of real concrete on breakwaters. This will occur slowly with time in a manner that is independent of weight and strength per se. Table 2 (from ref. 14) shows various features and effects on the durability of concrete with time. Fookes and Poole (ref. 14) give examples of deterioration of concrete with time, where in extreme cases, effective use of the concrete as armour has been lost in a couple of decades. Such concrete has been made with high w/c ratios (i.e. >0.55) and low cement contents (e.g. <300 kg/m^3) where perhaps the design strength requirement has been achieved but the durability may be lacking and some other factor, such as poor aggregate, may also be present. Concrete armour should therefore be designed and constructed with durability as well as weight and strength in mind.

A. H. BECKETT (*Sir Bruce White Wolfe Barry & Partners*)
Mr Baird has left us in doubt that the philosophy behind the rubble-mound breakwater is the acceptance of damage in extreme storm conditions. He quotes 14 papers on failure; at what stage does acceptable damage get renamed failure? He has thrown doubt on the Hudson formula which has enjoyed almost universal acceptance for design of armouring. This formula takes no account of wave period which seems to me to be wrong, for this is where the destructive energy of the breaking wave may be found. It is important for the following reasons:

(a) When a wave breaks a mass of water moves at the speed of the wave which is directly related to wave period. For example, the velocity of impact on a breakwater from an 18 s wave is three times greater than from a 6 s wave and the destructive energy is nine times greater.

(b) The sea bed profile limits the height of wave run up but not the period, so that we may have a wave height of say 3 m as the basis of design for armour.

(c) If the seaward slope of the breakwater is such that the vertical component of water movement exceeds the free fall velocity of the armouring in water, armour units will be floated out of position and subject to damage by falling when the wave subsides.

Taking our example of a 3 m high wave of 18 s period against a breakwater sloped at 1½ to 1, the vertical component of wave velocity is about 18 m/s, enough to cause splume to the height shown on the front page of the conference programme; the 6 s wave would produce a velocity of about 6 m/s which is insufficient rise above the spray wall. Yet both forms of breaking wave would merit the same armouring under Hudson.

May I suggest that tests be made to establish free fall velocities in water of all shapes and sizes of patent armouring with a view to finding the effect of long period wave action.

J. M. REBEL (*Hollandsche Beton-en Waterbouw B.V., Gouda*)
Paper 9 supports the contractor's point of view that one is much better off to bend to nature than to fight the elements. Mr Baird's design concept of a rubble-mound breakwater is governed by the quarry output and consequently the construction requires no special equipment.

Undisputably, two points are very much in favour of the contractor, and are thereby cost sensitive. Firstly, the contractor no longer needs to battle against nature to achieve slope angles steeper than the natural angle of repose of the material under the environmental loads they are subjected to. Secondly, the contractor is spared the strict tolerances imposed upon him by the use of sophisticated artificial armour blocks in the design.

In Mr Baird's concept the environmental forces work both for the designer and for the contractor.

Now, I am not certain if his suggestion is an indication that the designers of breakwaters should have more consideration for the constructability of their designs or that it is a hint at the desirability of 'Design and Construct' contracts with the quarry output as the major cost parameter.

P. H. KEMP (*University College, London*)
These comments relate to breakwaters in general, but they arise from the Authors' remarks that for rubble-mound slopes of about 1 in 5, material may be placed in the form of a mound and the desired slope produced by wave action. There is a direct connection between this method and the method of beach replenishment as practised, say, on the Kent coast (UK). The general point that arises is that breakwaters are devices designed to dissipate or absorb wave energy. Since work is a form of energy, and work = force x distance, one can regard a rigid vertically faced breakwater as one in which much of the energy is dissipated over a small distance with a correspondingly large force, and a gentle sandy beach as one where the run-up distance is large and the force, or friction on the beach is small. The latter situation is essential in the light of the small particle size of the sand. The rubble-mound breakwater comes in between these two cases, having some elements of the immobility of the rigid breakwater, together with the porosity of a beach. Clearly, the beach, with its ability to change its slope according to the severity of the wave attack and hence the distance over which the energy is absorbed, is by far the most effective and sophisticated wave absorber. It is of note that whereas on the basis of the Paper a rubble-mound breakwater designed with a slope 1 in 5 to withstand waves 5-6 m high would require elements of, say, 0.7-3.5 t in weight, beaches such as Chesil Beach at the Portland Bill and with a similar slope can dissipate waves of the same height with beach material of weight 1-2 kg.

P. ACKERS and J. D. PITT *(Paper 7)*
The comments of Dr Jensen are well appreciated and, in fact, an essential feature of our field research was a reliable survey method that was entirely compatible with the laboratory procedures. As the contributor says, the method should be tuned to the roughness of the surface being measured, and in our field work (as in the HRS laboratory research) the survey staff was provided with a hemispherical foot of diameter 0.5 D_{50}. Although rather difficult to manage in the largest size, especially on windy days, the method was entirely successful. Of course, in our case surveys were carried out 'in the dry': other problems would arise if one was surveying below water level, although similar principles should be followed if practicable.

No standard yet exists for the definition of the upper surface of slope protection, and hence of its thickness. As part of our study, we examined the variability of the surface by assessing the standard deviation of individual measurements about the mean surface.

Table 3 shows how the perceived thickness is influenced by the interpretation of the survey data. If only the highest 6% of levels are considered (assuming normal distribution), then the apparent layer thickness is increased by some 50% compared with the mean. This result depends on the survey technique: a 0.5 D_{50} foot to the survey staff is implicit and the result applies to riprap only. Dr Jensen's Fig. 1, relates to tetrapods and shows even greater variability using a foot much smaller than one half some effective armour size.

Mr Ligteringen questions underlayer porosity for the regular slope protection. There was no direct measurement of this but the method of construction was such that it had no bearing on the performance of the test panels. The filter material was somewhat variable but the grading was typically as follows: D_{10} 2 mm, D_{25} 10 mm, D_{50} 30 mm, D_{75} 50 mm, D_{90} 80 mm. The actual embankment was formed of a dredged silty sand of very low permeability.

The regular slope protection was modelled following the field-scale tests, alongside the test panels. Details will be found in ref. 11 of Paper 7. There was no discernible scale effect in these tests which may indicate that the underlayer modelling was valid. It is accepted that failure to model the underlay properly could give rise to spurious scale effects.

With regard to costs, it should be remembered that mobilization costs were not chargeable to the research budget and that much of the necessary instrumentation and data collected was funded in connection with the bank construction. The cost could be misleading therefore, but was of the order of £70,000 in the period 1976-77.

While sympathizing with the views expressed by Mr Lacey in Session 1, it is an oversimplification to suggest that the problems of breakwater design would be solved if the designer also supervised construction and post-construction monitoring. It is not generally true that 'the safety or assessment of failure risk, indicated by model tests, has not been borne out by prototype breakwaters, which have failed in both shallow and deep waters'. Some modelling deficiencies are recognized, for example the problem of modelling the structural strength of armour units which may fracture on displacement. It is perhaps understandable, though in some respects unfortunate, that very little attention is paid to successful projects. One example is the use of dolosse on the east sea cofferdam of the High Island project in Hong Kong with a design wave almost the same as the design wave at Sines. These projects were contemporary: one was entirely successful with no discernible damage over a ten year period. There are lessons to be learned from success as well as from failure and in the case of High Island the model testing, design procedures and construction have so far met requirements. Whether this would have been so had construction supervision been separated from design could be debated, of course. The problem of monitoring a secure project is to convince the client - or even oneself - that continuous wave measurements and surveys repeatedly showing no damage for year after year can be justified economically. It was this aspect that led to the construction of special test panels which would probably fail for the research described in Paper 7.

T. S. HEDGES *(Paper 8)*
Until recently, study of rubble-mound behaviour has been dominated by concern about armour stability, with little notice being taken of the fundamental roles played by the core and underlayers. In fact, it is somewhat ironic that a paper on the core and underlayers of a rubble-mound structure should be presented in a session on armour stability, for let us be clear: armour stability is an essential requirement for mound stability but it is not a sufficient requirement. For this reason, I am particularly grateful to Dr Barends for describing the work carried out at the Delft Soil Mechanics Laboratory on the geotechnical stability of rubble-mound breakwaters. While tests to assess the hydraulic stability of breakwaters are common, the work at Delft is the first comprehensive attempt to gain an insight into the geotechnical stability of the mound itself, as distinct from its foundation.

Dr Barends refers to the essential differences between the shearing behaviour of rockfill and of finer material such as sand. Under high effective stresses there is a significant degree of crushing of the contact points between stones in a mound. This crushing is important to geotechnical stability, as noted by Professor Burcharth, because it lowers the friction angle of the material. Barton and Kjaernsli (ref.15) provide a comprehensive discussion of this problem in relation to rockfill dams.

Mr Hunter draws attention to the different results obtained from using two methods of designing the top underlayer. Clearly, the modified filter rules method provides information only on the minimum size of material that may be placed in this layer. Coarser material may be necessary to ensure that the underlayer is satisfactory as a foundation for the armour, and to provide adequate temporary protection to the core before the armour is placed. Almost certainly, these aspects have been considered when preparing the guidelines in the US Army's Shore Protection Manual (ref.6), though they are not explicitly stated. I agree with Mr Hunter on the need for realistic tolerances when specifying underlayer thicknesses.

Mr Hall raises the question of whether geotextiles could be used to reinforce a breakwater core or its foundation. Undoubtedly, this is possible as there are many examples of geotextiles used in a similar way on land. However, as mentioned in Paper 8, the durability of a potential construction material is one of the primary properties to be considered. At present there is little information on the durability of geotextiles in the marine environment. This is a subject which deserves further research.

I am grateful for the contributions of Mr Harlow, Dr Barber and Mr Maquet. I agree with Mr Maquet that the question 'What type of armour will be placed over the underlayer?' should be added to the list in paragraph 33 of Paper 8.

In Session 1, Mr Lacey questions the wisdom of splitting the design, supervision of construction, and monitoring processes between various bodies. When such arrangements apply, they may compound the problems, but I do not believe that they are the root cause. Rather, I would turn to something else which Mr Lacey says: '... our studies have persuaded us to steepen the breakwater slopes to excess, aided by the apparent properties of various man-made armour units'. Almost certainly, many problems have resulted from the preoccupation of designers with the armour layer. This preoccupation has caused them to overlook potential failure mechanisms which are not linked to the armour, such as damage to the back-slope by throughflow and slumping of the mound resulting from a loss of shear strength during rapid drawdown (see Figs 4 and 5 of Paper 8). These problems have been aggravated by lack of understanding of the limitations of short-term wave measurement programmes and of physical model studies.

If we had 1000 years of wave observations at a site then waves with an average return period of 50 or 100 years could be determined fairly accurately. But instead we may have only one or two years of records and extrapolation to 50 or 100 years will not provide very reliable estimates. Is the design engineer aware of how unreliable his estimates are (see ref.16)? Furthermore, tests on model breakwaters are valuable but they can never totally reproduce (or predict) the behaviour of a full-scale structure. They must never be used by the designer as a substitute for thought, but only as an aid to thinking. Model tests must be supported by calculations of the sort described by Dr Barends. Even elementary calculations can help to provide the engineer with what Mr Lacey calls a 'gut engineering' feeling for his problem. For example, stability computa-

tions for seepage down an infinitely long sand slope (see ref.17) suggest that the slope is subject to shear failure if it is at an angle greater than about $\emptyset/2$ ($\emptyset$ is the angle of friction of the material). Is the situation changed significantly by covering the slope with underlayers and armour? If not, then we should be very cautious about using breakwater slopes steeper than 1 in 2.

REFERENCES
1. BRUUN, P. and KJELSTRUP, Sv. Design of mound breakwaters, 1983.
2. BARENDS, F.B.J. and THABET. Groundwater flow and dynamic gradients, 1978.
3. HANNOURA and BARENDS, F.B.J. Numerical modelling of rubble-mound breakwaters. Proc. 4th Conf. FE in Water Res., Hannover, 1982.
4. SWEET et al. A method for dynamic soil structure interaction problems. Proc. Conf. Numerical Methods for Non-linear Problems, Swansea, 1980.
5. BARENDS, F.B.J. et al. West breakwater – Sines – Dynamic geotechnical stability of breakwaters, Proc. Conf. Coastal Structures '83, Arlington, 1983.
6. US Army Coastal Engineering Research Centre. Shore Protection Manual. 3rd edition. US Government Printing Office, Washington, 1977.
7. THOMPSON, D.M. and SHUTTLER, R.M. Design of slope protection against wind wave attack, CIRIA Report 61, 1976.
8. GROENVELD and JENSEN. Coastal Structures '83, Arlington, 1983.
9. Concrete armour units for protection against wave attack. US Army Hydraulics Laboratory miscellaneous paper H-74-2, January 1974.
10. An appraisal of rubble-mound breakwater. Dock and Harbour Authority, November 1981.
11. Stability of dolos blocks under oblique wave attack. Hydraulics Research Station report IT 159, April 1977.
12. Forces on sea walls under oblique wave attack. Hydraulics Research Station report IT 225, July 1982.
13. THOMAS STEVENSON. Design and construction of harbours. Adam and Charles Black, Edinburgh, 1874.
14. FOOKES, P.G. and POOLE, A.B. Some preliminary considerations on the selection and durability of rock and concrete materials for breakwaters and coastal protection works. Q.J. Eng. Geol. 14, 1981, 97-128.
15. BARTON, N. and KJAERNSLI, B. Shear strength of rockfill. J. Geotech. Engng Div. Am. Soc. Civ. Engrs 107, No. GT7, 1981, 873-891.
16. WANG, S. and LE MEHAUTE, B. Duration of measurements and long-term wave statistics. J. Waterway, Port, Coastal and Ocean Engng Div. Am. Soc. Civ. Engrs 109, No. 2, 1983, 236-249.
17. LAMBE, T.W. and WHITMAN, R.V. Soil Mechanics, S.I. version. Wiley, New York, 1978.

10 Risk analysis in breakwater design

A. MOL, H. LIGTERINGEN and A. PAAPE, Delft Hydraulics Laboratory

A general outline is presented of the applicability of probabilistic design to rubble mound break-waters. For such a structure the possible causes of failure are schematized into a faulttree. After having described various computational levels of approach, the results are given of a case study, related to the West Breakwater at Sines. It is shown to what extent various design aspects contribute to the overall risk of failure.

1. INTRODUCTION

Until now the design of rubble mound breakwaters was generally based upon a combination of experience, engineering skill and hydraulic model-studies to define the wave climate and the hydraulic stability. The criteria for the design were set by defining the return period of storm conditions (50-100 years) during which beginning of damage might occur. These methods and criteria used in the design process should introduce a sufficient safety margin between load and resistance to prevent severe damage or collapse of the breakwater.

However, in recent years a number of large breakwaters suffered severe damage. From studies of these failures it was concluded that damage was caused by a combination of aspects (wave climate, structural strength of armour, geotechnical stability, constructability). The safety margin for these structures was not large enough and it may be concluded that the traditional design approach was inadequate. In probabilistic terms, taking into account all stochastic variables influencing load and resistance, the probability of failure was too high, causing a high encounter probability of a certain damage in the years after completion of the breakwaters.

A certain risk of failure in the lifetime of the breakwater always exists, due to the stochastic character of load and resistance. Ideally the probability of failure should be fully quantified in the design process. Criteria for the encounter probability of a certain damage should be established taking a number of aspects into account.

. economy

The total cost of the structure consists of construction costs and the anticipated maintenance and repair cost. From economic point of view the optimum design has the lowest total of construction cost and capitalized repair cost.

. safety

A construction can be of vital importance for a larger system (for example the breakwater for the functioning of the harbour or berths behind it). In this case the probability of failure has to be based on the accepted risk of a similar breakdown in the harbour operations, due to other causes.

other aspects

Political or social aspects may sometimes influence the probability level of a certain failure. Also the estimated future effort on inspection and maintenance during its lifetime will affect the accepted probability of failure.

In an earlier paper by Van de Kreeke and Paape (ref.1), describing methods of economic optimization of breakwater design, a quasi-probabilistic approach was used. This implied that only the wave height was considered to be stochastic, whereas the other variables describing load or strength were treated deterministic. Also only one failure mode (hydraulic damage to the armour layer) was taken into account.

In civil engineering more sophisticated probabilistic methods are at present rather in a state of development. The understanding has developed that a purely deterministic approach is unsatisfactory from an engineering standpoint (safety factors rely on empirism and do not allow large extrapolations beyond the field of experience) and from an economic point (the design is often conservative). On the other hand a probabilistic approach requires a far better understanding of the physics and mechanics involved in the design then is often available. Much effort has therefore to be spent on improving this knowledge. In this respect studies on collapsed breakwaters are of great importance to understand the failure mechanism.

In this paper the probabilistic approach is further elaborated for the case of a rubble mound breakwater, integrating the statistical information on several important variables (wave climate, stability, concrete strength) into the probability of classes of damage up to failure. Apart from quantifying thus the risk of failure, it is shown how the probabilistic approach can serve the designer in achieving a

Breakwaters—design and construction. Thomas Telford Ltd, London, 1984

133

balanced study approach in which most effort is put into those parameters that have the greatest contribution to the total probability of failure.

2. THE PROBABILISTIC APPROACH

To analyse the risk of failure of a structure sufficient information about the system should be collected. All stochastic variables influencing strength and load have to be described by their probability density function (pdf). The failure mode of each component of the structure has to be investigated. An important stage in the probabilistic design is the set-up of a faulttree, comprising all possible failure modes of various components of the structure, each with their own partial probability of failure. Three methods may be distinguished of determining the probability of failure of a system (ref.2), which are listed in order of accuracy and complexity as follows:

Level I quasi-probabilistic approach

For each of the stochastic variables a certain unfavourable value is chosen, the so-called characteristic value. Further a set of partial safety factors is applied, establishing the margin between the extreme load and the average strength. When L_c is the characteristic load, γ_s a safety factor and R_c the characteristic resistance, the failure function is written as

$$F = \gamma_s R_c - L_c \qquad (1)$$

and failure occurs for $F \leqslant 0$. This method corresponds best with the traditional design method described in Chapter 1.

Level II probabilistic approach

The load L and resistance R are now treated as stochastic variables and the failure function reads:

$$F = R - L \qquad (2)$$

The failure function at this level is approximated by linearization. Several methods have been developed in this regard, which may be described in suborder of accuracy and complexity:
. mean value approach, linearization by using the mean values in the failure function
. advanced first order-second moment approach, linearization of failure function in its point with maximum probability density (design point)
Both methods assume that all stochastic variables follow the normal distribution characterized by mean value μ and standard deviation σ. This is further improved in the third method:
. advanced full distribution approach (AFDA), where the actual distribution is approximated by a normal distribution with the same probability density in the design point
To put this in algebraic terms the failure function is written as:

$$F = F(x_1, x_2, \ldots x_n) = 0 \qquad (3)$$

in which x_i are the stochastic independent variables. If X_i defines the above mentioned design point a first approximation of the linearized failure function in this point is given as follows:

$$F = F(X_1, X_2 \ldots X_n) + \sum_{i=1}^{n} \frac{\delta F}{\delta x_i}(x_i - X_i) \qquad (4)$$

In the advanced first order-second moment approach the mean value and standard deviation read:

$$\mu_F = F(X_1, X_2 \ldots X_n) + \sum_{i=1}^{n} \frac{\delta F}{\delta x_i} (\mu_{x_i} - X_i) \qquad (5)$$

$$\sigma_F = \left[\sum_{i=1}^{n} \left(\frac{\delta F}{\delta x_i}\right)^2 \sigma_{x_i}^2 \right]^{1/2} \qquad (6)$$

Solving $F = 0$ provides a new estimate of X:

$$X_i = \mu_{x_i} - \frac{\delta F}{\delta x_i} \cdot \frac{\sigma_{x_i}}{\sigma_F} \cdot \mu_F \qquad (7)$$

after which the computation is iterated until sufficient accuracy is obtained.

Since in this approach F follows the normal distribution as well, the probability of failure P_F may be written as

$$P_F = 1 - \text{erf}(\beta) \qquad (8)$$

in which (see also Fig.1):

$$\beta = \frac{\mu_F}{\sigma_F} \qquad (9)$$

The value of P_F can be found from the tables of the normal distribution.

Although the level II approach is less accurate than the next level, it has the advantage of being easily applicable and providing insight into the contribution of various variables to the overall probability of failure. Where the mean value approach in most cases still allows analytical treatment, the latter two methods at level II require already a computer.

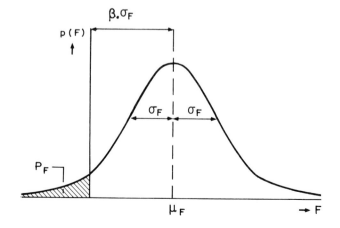

Fig.1 Probability density function of F, definition of reliability index β.

Level III full probabilistic approach

At this level an exact calculation of the probability of failure is made by convolution of the probability density functions of the load L and resistance R:

$$P(F<0) = {}_o\int^{\infty} f_r(\emptyset) \ F_1(\emptyset) \ d\emptyset \qquad (10)$$

$f_r(\emptyset)$ = probability density function of the resistance R.

$F_1(\emptyset)$ = cumulative probability distribution function of the load L.

Where the functions become too complex for direct computations the use of Monte-Carlo simulations provides an alternative. In general however these methods require more computational effort than warranted in view of the knowledge of the probability density functions.

3. APPLICATION IN BREAKWATER DESIGN

In Figure 2 a main faulttree of a rubble mound breakwater is shown schematically (not necessarily complete). The final failure of the breakwater is assumed to be its disappearance below mean sea level by removal of the capping wall. The loads on the structure are wave exposure and seismic activities. These loads may cause events such as hydraulic instability of front and rear armour, geotechnical instability of front and rear slope, or a destruction or sliding of the capping wall, direct or as a consequence of the damage to the slope. For each event the (partial) probability density function should be defined. Some events may be independent, while other events are dependent. It will be clear that the total probability of failure of the breakwater is greater than the probability of hydraulic instability of the front armour layer only. In a proper design the probabilities of different events, each leading to total failure of the structure, should be balanced. In the present case only the event "damage to seaward armour layer" due to the event "wave exposure" is treated.

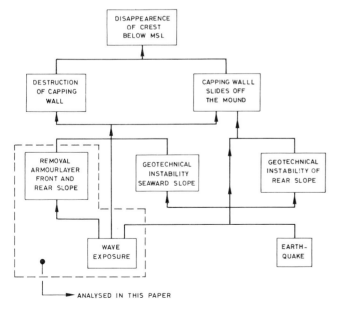

Fig.2 Main faulttree of a rubble mound breakwater.

Referring to Chapter 2 a sufficient insight in the failure mechanism should be present, in order to determine the contribution of all stochastic variables to the overall probability of failure. However, for breakwaters a general quantitative description of the relation between damage and wave condition is not yet available. A heuristic description is given by for instance the Hudson formula:

$$W = \frac{\rho_c \ g \ H^3}{K_D \left(\frac{\rho_c}{\rho_w} -1\right)^3 \cot g \alpha} \qquad (11)$$

with W = weight of armour unit
H = wave height
ρ_c = density of concrete
ρ_w = density of water
α = slope angle

This formula does not describe the influence of variables such as foreshore, toedepth, crestheight wave period and groupiness. The shape (and roughness) of the armour are represented in the K_D-factor. It is generally accepted that this formula can only be used in an indicative way. For a certain design the actual relation between wave conditions and damage should be established by hydraulic modeltesting.

Level I approach

At present a breakwater is generally designed in accordance with a level I approach. For the characteristic load generally a waveheight with a return period of 50 or 100 years is selected. The probable range of peakperiods and/or groupiness is assessed and the effect is checked in a wave flume. An average or unfavourable period or groupiness is then selected for the characteristic load. The criterium for the design is as described in the introduction. In this approach the effect of uncertainty in the results of wave climate, modeltesting and construction cannot be shown.

Level II approach

This approach to analyse the risk of failure of an armour layer was proposed by Agema and Vrijling (ref.3), using only the Hudson formula for the relation between load and resistance. In the following the failure function F is derived, using also the load and resistance in terms of wave height, but including the uncertainty in the result of the wave climate study, in refraction effects and in the results of hydraulic modeltesting. This is carried out by calculating the K_D-factor with the formula of Hudson, using the wave height at which a specified damage in the wave flume occurred. It is further assumed that the Hudson formula describes the influence of the parameters W, ρ_c, α correctly. The failure function F may then be written as follows:

$$F = R - L \qquad (12)$$

with: $L = \alpha_w H_s$

$R = \frac{(\rho_c/\rho_w - 1)}{K_r(T_p, \emptyset)} \left(\frac{W.K_D.C(T_p, \emptyset) \cot g \alpha}{\rho_c \ g}\right)^{1/3}$

$K_r(T_p, \emptyset)$: refraction function
T_p : peak period wave spectrum
$\emptyset$: wave direction
$C(T_p, \emptyset)$: correction function on K_D-factor

The value of $K_r(T_p,\emptyset)$ should be obtained from a refraction study, while the K_D and $C(T_p,\emptyset)$ should be calculated from stability tests in a hydraulic model.

The following problems are faced when using a level II approach:

All stochastic parameters are assumed to be normally distributed. This is never the case and approximations should be made. This can be done by fitting a normal distribution to the real distribution around the failure-point.

All parameters are assumed to be statistically independent. This is not the case for the wave-parameters H_s, T_p and $\emptyset$ which will show often a certain dependency. To deal with the problem, the distribution for T_p and $\emptyset$ can be used, as found in the most probable failure-point.

Level III approach

The actual probability functions of L and R, defined in the level II approach, should be determined. This is relatively simple for the load, but more complicated, although not impossible, for the strength. By convolution the probability density function of F is obtained. This approach will give undoubtedly the most accurate value of the probability of a certain damage, while all types of distribution functions can be used. As will be shown, it gives no insight in the contribution of a variable to the overall variance of the failure function, as is obtained in the level II approach.

4. CASE STUDY ON SINES BREAKWATER

A well known recent failure is the damage on the West Breakwater of the Port of Sines in Portugal. An overview of the original design process, the criteria used therein and the most probable failure mechanism is described in a report of a ASCE commision (ref.4). In studies carried out by the Delft Hydraulics Laboratory for the repair, a new wave climatology was defined for the Sines area (ref.5).The data presented in references 4 and 5 will be used in the following to calculate the risk of failure by a level II approximation, using the advanced first order-second moment approach. The calculations are made by computer. The aim of this risk calculation is to demonstrate the contribution to failure of different variables and to emphasize the importance of a thorough study on the wave climate and stability, since a reduction of the uncertainty in these factors is directly translated into a more economic design (in the probabilistic approach). Figure 3 gives the relation of damage and probability of occurrence (defined again as return period) when using the original design criteria either in a level I approach or in a probabilistic calculation with one or more variables treated stochastically. Further this relation is calculated on the basis of the new design criteria and presented in Figure 3.

A. Original design criteria

The design criteria for the breakwater were based upon the presumption that repair works should not affect the harbour operations too much. A damage of 1 % was permitted for wave conditions with a return period of 100 years and first damage might

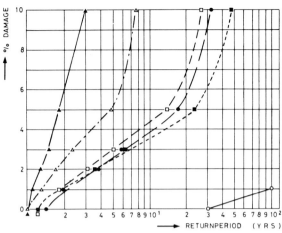

A - QUASI PROBABILISTIC DESIGN
B - ORIGINAL WAVE CLIMATE, HYDRAULIC DAMAGE
C - ORIGINAL WAVE CLIMATE WITH CONFIDENCE BAND, HYDR. DAMAGE
D - ORIGINAL WAVE CLIMATE INCL. REFRACTION, HYDR. DAMAGE
E - NEW WAVE CLIMATE INCL. REFRACTION, HYDR. DAMAGE
F - NEW WAVE CLIMATE, HYDR. AND STRUCTURAL DAMAGE

Figure 3 Return period of damage West Breakwater, Sines

occur for conditions with a return period of 30 years. The wave climate study was at that time based on observations at Sines (a few years) and at a location 200 miles North of Sines (over 7 years). These sources gave a considerable difference in the 100 year wave height and by averaging a value of $H_s = 11$ m was derived.

B. Hydraulic stability as stochastic variable

Very short stability tests with random waves were carried out before construction started, aimed at measurement of overtopping. These results are therefore not used here.

Stability tests, carried out after the failure at several laboratories, showed a different relation between damage and wave height (Figure 4). With unbroken dolos the 1 % damage was reached at $H_s = 7$ m. At $H_s = 11$ m a severe damage was found. Using the data from stability tests and the wave climate as originally established the risk of failure is shown in Figure 3. A considerable decrease of the return period is found.

C. Uncertainty of design wave height (deep water)

When defining the return periods of wave heights in deep water, a large difference was found between two sources. As mentioned before, the final assessment was made by averaging the results. The risk of damage is calculated now by considering a confidence band α_w around the wave height exceedance curve with a standard deviation $\sigma/\mu = 0.15$. When assuming deviations from the mean of 2σ the actual design wave height can therefore vary between 8 and 14 m. The resulting risk damage relation is also shown in Figure 3. The return period of a given damage is reduced by 50 % compared to B.

D. Refraction effects

In the original wave climate study no refraction effects were expected. In reference 5 it is shown

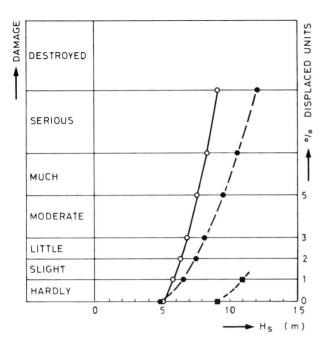

- ■ DESIGN CRITERIA
- ● TESTRESULTS INTACT DOLOS
- ○ TESTRESULTS BROKEN DOLOS (ESTIMATED)

Figure 4 Hydraulic damage versus wave height

that the refraction coefficient is a function of the wave period and direction, and has an average value of 0.95. Using the refraction coefficients in the risk calculations it is shown that the return period increased.

E. Implication of the new wave climatology
In reference 5 the results of a recent hindcast study for the Atlantic Ocean are presented. The design wave height for the West Breakwater at Sines was accordingly adjusted to H_s = 14 m. Some scatter still occurs when using different extrapolation techniques. In the calculations σ/μ = 0.05 is used. A considerable decrease in return period is found in Figure 3.

F. Breakage of dolos
In tests carried out at the NRC laboratory in Canada (ref.4) it was found that breakage of dolos affected to a large extent the stability of the armour layer. From these results a more realistic damage function for the 40 t dolos was estimated (Figure 4). The results of the probabilistic calculations was that severe damage will occur for wave conditions with a return period of 3 years and an encounter probability of 0.33 per year.

A different method to show the effect of the stochastic variables on the total probability of the failure is presented in Table 1. The contribution of each parameter to the overall variance σF^2, expressed in

$$\left(\frac{dF}{dx} \frac{\sigma_x}{\sigma_F}\right)^2 \qquad (13)$$

is presented for the computational cases D and E.

Situation D

Parameter X	μ	σ	$\left(\dfrac{dF}{dx} \dfrac{\sigma_x}{\sigma_F}\right)^2$
T_p (s)	18	2.5	0.08
$\emptyset$ ($^\circ$)	0	15	0.02
α_w	1.0	0.15	0.27
H_s	6.4	1.3	0.63

Situation E

T_p	18	2.5	0.09
$\emptyset$	0	15	0.01
α_w	1.0	0.05	0.04
H_s	8.1	1.3	0.86

Table 1 Constitution of the variance σ_F^2

In the Table it is shown that the risk of failure is not only gouverned by the probability of a certain wave height, but also by the wave period and by the uncertainty in the prediction of the wave height. The wave direction showed (in this case) hardly any effect. By using only mean values in the design, and overlooking the effect of stochastic variables other than the wave height, the probability of a certain damage will be underestimated. Hence a design tested with average values of e.g. period and direction (but also other parameters influencing strength as e.g. density and weight) may show an acceptable risk in a level I approach, but an unacceptable risk in the level II approach. To counteract this effect in the level I approach a maximization is achieved by using more extreme conditions: in fact an extra safety margin is introduced. The real probability of failure is not known however, which may lead to a too costly, or still too risky design.

5. CONCLUSIONS
The total probability of failure of a breakwater is composed of the partial probabilities of failure with respect to hydraulic instability, geotechnical instability and subsidence of the capping wall due to external loads such as wave exposure and seismic activities.
The example provided in this paper is still limited to the hydraulic stability of the seaward armour layer.

The present design method of breakwaters is based on a quasi-probabilistic approach (level I): a characteristic load is established (a wave height with a certain return period) at which hardly any damage should occur. The real probability of failure is then not known due to the uncertainties of other parameters affecting stability.

To establish a better approximation of the risk of a certain degree of damage a probabilistic approach (level II) is presented, assuming a normal distribution of all stochastic variables influencing load and strength. This method was used to analyse the risk of failure of Sines

West Breakwater with various assumptions concerning stability and wave climate.

It is shown that the commonly used level I approach, designing a breakwater with only one stochastic variable (the wave height), may result in an underestimation of the return period of the considered damage, due to the contribution of other variables influencing the stability. Hence when a certain return period is required a safety margin should be applied by using more infavourable values of the stochastic variables. The magnitude of the safety margin will depend on the uncertainty of the other variables.

A level II approach gives more realistic approximation of the risk of failure, and gives also an insight in the contribution of all stochastic variables involved. Such an insight may help to define the priorities in the design-studies leading to a balanced approach.

The case study carried out for Sines West Breakwater has shown once more the importance of accurate wave climate studies and hydraulic modeltests for a minimization of the risk of failure. The level II approach enables to transform uncertainties and inaccuracies in the studies to the risk of failure.

REFERENCES
1. KREEKE, J. VAN DE and PAAPE, A, On optimum breakwater design, 9th Conf. on Coastal Eng., Lisbon, 1964.
2. RACKWITZ, R, First order reliability criteria Appendix 1, Vol.1, Bulletin d'Information N 124/125, CEB, Paris, 1978.
3. AGEMA, J.F. and VRIJLING, J.K., Concrete armour units and probabilistic design of breakwaters. Postdoctorate course on Harbours. Delft University of Technology, 1981 (in Dutch).
4. Port Sines Investigation Panel. Failure of the breakwater at Port Sines, Portugal. ASCE-report, New York, 1982.
5. MYNETT, A.E., VOOGD, W.J.P.de, SCHMELTZ, E.J., Wave climatology Sines West Breakwater, Coastal Structures Conf., Washington, 1983.

Discussion on Paper 10

P. LACEY (*Ove Arup & Partners*)
Chairman's Introduction to Session 5
In 1977 at the Isle of Wight symposium on rubble-mound breakwaters very little was said about risk and risk analysis apart from a paper by Dupuy & Brennan which caused a stir by discussing design risk fixing by the client, and one by Per Bruun & Gunbak who looked at risk criteria in the design stability of sloping structures. Now some six years and several failures later, there is still considerable more attention being paid to risk. The client and design team have to address themselves to risk that data is inadequate, risk that interpretation of it is suspect, risk that tests are not realistic and risk that construction is not up to standard. It is interesting that Mr Baird (Paper 9) has reinvented the Rennie method, as I feel our greed and that of our clients has led us to increase slope steepness beyond our technology and therefore we are at risk.

The Authors of Paper 10 have tried to assess the methods for judging the risk of failure.

A. MOL, H. LIGTERINGEN and A. PAAPE (*Introduction to Paper 10*)
Paper 10 presents a still very limited application of a probabilistic method to a case study: of all the possible failure mechanisms shown in Fig. 2, we confine ourselves to the effects of waves on the armour layer. And within that limited scope the stability of armour units is treated in a very global way, by using the Hudson formula.

Yet this application already provides some very interesting results, as we have tried to indicate in Fig 3. We have used the case of the original design of Sines, since so much data are available on that breakwater. The conclusions of this exercise are that the risk of failure depends very much on the method of model-testing and their interpretation, the definition of the wave climate, and the inclusion of the strength aspects in the evaluation.

These factors are not surprising but what is interesting is the extent of their influence. Taking this into account and realizing that this is only a limited part of the overall risk analysis (geotechnical aspects may further increase the risk of failure) one may wonder why so many breakwaters are still functioning properly.

I have two comments to make:
- many breakwaters have been designed and built within the limits of past experience,
- even when breakwaters are functioning properly,

damage and repair is a continuous cycle for many of them, be it less spectacular than the most recent cases.

The work described in this Paper should be expanded in two directions:
- to include more relevant failure mechanisms
- to improve the determination of probability density functions for each of the failure mechanisms.

Whereas the first is a matter of applying the existing knowledge (and can therefore be taken up in the design process), the latter will form a subject of further research by hydraulic, geotechnical and concrete institutes.

In this respect my plea is to concentrate the research on developing mathematical-physical models of processes in the faulttree. This should be an integrated effort, countrywise and by discipline.

In the short term the available tools should be put to good use. For example, computational techniques have gradually allowed the assessment of the geotechnical stability of core underlayers and toe. Relations between block type and size and allowable impact velocities have been determined, as Professor Cusens showed in Paper 4. By acceleration measurements in the model laboratories can indicate what percentage of units exceeds a given impact-velocity for a given wave condition. Insight in scale effects has increased considerably during the last few years. The development and validation of numerical hindcast models facilitates the definition of wave climate, taking their inherent limitations into account for virtually any location in the world.

I would like to finalize this introduction with an example of our present attempts to study the damage development on the armour layer. Table 1 gives details and results of tests to determine the behaviour of armour units under water. From these results we try to obtain quantitative descriptions of the loading cases for rock, tetrapods and dolos.

M. A. MESTA (*Sezai Turkes-Feyzi Akkaya Co. Inc., seconded to Massachusetts Institute of Technology*)
No matter which design method is used, one must be aware of

(a) the strength of the structure

(b) the influence of the construction methods on the final strength of the structure

(c) the stochastic character of the boundary conditions.

I will concentrate on the stochastic character of the boundary conditions.

There is no common feeling about in what form the boundary conditions should be formulated. Furthermore, it is not a clearly stipulated fact on what criteria the use of a certain type of structure or structural element should be based. Both the formulation of boundary conditions and the criteria for selection of type are strongly related to the behaviour of the structure, the prototype structure, in the environment it would be subjected to.

For instance, the presentation of the boundary conditions, in principle, is determined by the response properties of the structure for which the observation of the responses of a prototype is emphasized.

When both water levels and waves are important, and the structure reacts linearly on waves, the two-dimensional probability distribution function of the wave spectra and sea levels would be the most appropriate form to present the boundary conditions.

If the structure reacts non-linearly, which usually is the case, waves have to be given as combinations of both wave heights and periods as well as the water levels.

Safety should at least be defined within specified lower and upper confidence limits. There are three principal approaches to the assessment of structural safety.

(a) Traditional safety factor approach
This is simply the ratio of load at failure to working load, which does not admit the possibility of loading combination exceeding the strength, thus resulting in failure.

(b) Probabilistic approach
This accounts for both the variation in strength (the capability to withstand loading) and the variation of maximum loading (the demand placed on the structure), where it is practically difficult to determine accurately the probability distribution for the 'capability' as well as the 'demand'.

(c) Semi-probabilistic approach
This makes some allowance for the uncertainty concerning the maximum load to be placed on the structure and also for the uncertainty concerning the exact strength of the structure.

Figure 1 shows that a characteristic load may be defined or determined which has a 5% probabi-

lity of being exceeded, and a characteristic strength can be found for which the structure has 95% probability of exceeding.

Thus

$$D_K = \overline{D} + K_D \cdot \sigma_D$$

$$C_K = \overline{C} - K_C \cdot \sigma_C$$

where $\overline{D}$ and $\overline{C}$ are mean values and σ_D and σ_C are standard deviations.

Assuming demand and capability curves to have approximately normal distribution one would have $K_C = K_D = 1.65$, and allowing C_K and D_K to give C_K/D_K = in terms of partial safety factors, thus a safety index β_F, similar to the approach in Paper 10 could be employed:

$$\text{Safety index} = \beta_F = \frac{\overline{C} - \overline{D}}{\sqrt{\sigma_C^2 - \sigma_D^2}}$$

With reference to the definition of risk being a multiple of 'event' and 'consequences', I wish to emphasize that these factors are mostly involved in construction or its consequences. This again indicates the essence of incorporating construction within the design phase.

Several authors and speakers have referred to uncertainty and risk, which I wish to define.

(a) Uncertainty refers to situations where there is no suitable past data on which to estimate the chance of an occurrence.

(b) Risk pertains to situations for which there is previous probability data useful in predicting outcomes.

However, I would still refer to the statement that 'risk is an exposure to economic loss arising from involvement in the construction process', pointing again to the importance of the involvement of construction, via the risk process, into the design process.

The designer/consultant should realize and announce the role of instrumentation in prototypes in order to convince the client (owner) of the necessity and importance of them at an early stage of a project as is possible; much before the maintenance (monitoring) phase, for better assessment of the safety or risk lies in the prototype.

Furthermore, it is clear that this leads to the necessity of a systematic design approach, analysing the failure mechanism of the designed - built - prototype structures, in which some risk

Table 1. Behaviour of armour units under water*

Test no.	Armour protection		H_{s_i} :m	T_p :s	Damage after 15 min:%
1	Rock	317 g	0.236	2.73	2.4
2 a	Tetrapod	208 g	0.179	2.34	< 1
2 b	Tetrapod	208 g	0.25	2.77	destroyed
3 a	Dolos	162 g	0.24	2.75	1.6
3 b	Dolos	162 g	0.25	2.92	20

*Water depth d = 0.80 m; All slopes α = 1.2; In all tests approximately the same breakertype was maintained (surf similarity parameter ~ 3.5)

analysis techniques would be very helpful in the
failure analysis of a structure.

J. D. METTAM (Bertlin & Partners)
In para. 4B of Paper 10 the Author implies that
the random wave hydraulic model tests for the
original design dealt only with overtopping.
This is not correct. Random wave tests were
completed, after construction started but before
the main cross-section was reached. The final
tests showed only 1% damage with H_s = 11 m.
    The design was modified on the basis of those
tests and issued for construction without waiting
for the model test report. Despite requests by
the consultants, no report was received until
some time after the accident, when a report was
received which described only the overtopping
tests.
    In his opening remarks the Author commented
on a suggestion made during an earlier discussion
that specific gravities in flume tests should be
reduced, so that we know on completion of tests
that there is a certain factor of safety against
the movements or failure condition shown in the
model tests.
    His method of risk analysis is exactly compa-
rable to limit state design as used in this
country for structural design. Characteristic
loads and characteristic strengths are
established on a probabilistic basis. In using
these in design calculations partial factors
of safety are applied which may be only 1.0
for structural dead loads but are larger for
less certain loadings.
    However careful and sophisticated our
wave analysis and modelling techniques are, it
is still advisable to include some factor of
safety between our final model tests and our
completed design.
    This could be done as Clifford (theme paper:
the design process) suggests by using an
increased wave height in the model - but this
does not work in a depth-limited situation and
in deep water it distorts the geometry of wave
attack on the structure.
    It could also be done by arbitrarily
increasing armour weight in the prototype - but
this also alters the geometry of the structure
and may invalidate tests on overtopping and
drawdown.
    By analogy with structural design one might
use a fluid of greater density than water to
apply increased loads in the model - but there
are many problems in finding a suitable liquid.
    Much simpler, and completely effective, is a
reduction in specific gravity of the important
elements of the breakwater, decreasing the
stability of the model breakwater in the ratio
of the model breakwater in the ratio of submerged specific weights.

W. F. BAIRD (W. F. Baird & Associates, Ottawa)
Could the Authors describe the motions, settlements, etc. when they would expect armour units
in prototype to break?

H. LIGTERINGEN (Delft Hydraulics Laboratory)
I would like to comment upon suggestions made at
this conference. Mr Clifford indicated the

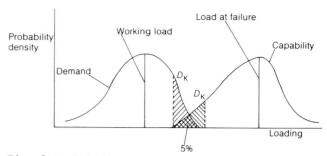

Fig. 1 Probability density against loading

possibility of using H_{max} in design and Mr Mettam
suggested a lower specific weight of the model
armour units than scale law would require. In
both cases I would advise against these suggestions, since those approaches tend to obscure
the actual risk of a design, rather than contribute to a better quantification.

A. MOL, H. LIGTERINGEN and A. PAAPE (Paper 10)
I have no specific comments regarding Mr Mesta's
discussion. His formulation of the safety index
appears to be equal to our reliability index β
for the special case that the failure function
can be described by two stochastic parameters
only, which both follow the normal distribution
approximately. Therefore, we agree with Mr
Mesta in his observation that the construction
process has to be taken into account and that
the monitoring of completed structures will
greatly contribute to a better understanding of
the failure mechanisms.
    I agree with Mr Mettam when he states that
the random wave tests were completed after
construction started, but before the main cross-
section was reached. This alters our wording
in para. 4B, but not the conclusion. Our understanding of the way model tests were incorporated
in the original design process is that the hydraulic stability was investigated in tests with
monochromatic waves. At a later stage overtopping was measured during short tests, applying random waves. These tests were not aimed
at providing adequate information on the
stability and therefore these results were not
used in this Paper.
    With regard to Mr Mettam's second point, we
would like to make the following comments.
Paper 10 demonstrates the advantages of a level
II or III probabilistic approach above the use
of characteristic loads and strengths with
partial factors of safety. The advantage lies
in the improved quantification of the risk and
the fact that contributions of various parameters
to the overall risk are identified.
    At right angles to these advantages works
the suggested reduction of block weights in
model tests: it reduces the risk (to what extent
remains unclear) but the relative importance of
the armour stability in the whole structure
becomes distorted. We would strongly discourage
this line of thought.
    Naturally Mr Baird is right in that my intro-
duction provides only reliable information
regarding the beginning of motion in the armour
layer: as soon as breakage occurs at full scale
the development of the damage will be different.

11 Construction methods and planning

J. F. MAQUET, IPC, Sofremer, Paris

After reviewing the main problems of the Construction of a rubble-mound breakwater, Construction methods are described considering use of floating or land-based equipment. The relation between design and construction is emphasized.

INTRODUCTION

1. Construction of a breakwater means to supply and to place suitable materials, with appropriate equipments, at a location exposed to the action of waves, according to a prepared design. If Construction is based on a design, it is clear that the construction methods and planning should be integrated with the design.

2. Bearing in mind the relation between design and construction, this paper will deal with construction methods and planning of rubble-mound breakwaters with special reference to the findings adopted by the 3rd Waves Commission of PIANC (ref.1).

3. The major problems encountered when planning the construction of a rubble-mound breakwater come from :

- environmental conditions during the construction
- availability of materials from the quarries
- working of a large volume of concrete
- availability of equipment.

Construction methods are :

- construction by floating equipment
- construction by land-based equipment.

ENVIRONMENTAL CONDITIONS DURING THE CONSTRUCTION

Statistical wave analysis

4. Choice of structure and selection of construction procedures highly depend upon construction climate, i.e. the wave conditions every day, month after month, season after season. This requires statistical wave analysis so as to evaluate, on an annual basis, the number of days and the duration of the longest period during which the weather conditions are judged to be unfavourable with respect to operation of the equipment as well as to the stability of the structure while construction works proceed.

5. A too optimistic evaluation of construction climate may induce longer construction time, inadequacy of selected construction equipment, modifications in the design to ensure stability.

Forecasts during construction

6. A good statistical wave analysis is not enough to plan the construction. It is necessary to organize the progress of the work according to the forecast weather sufficiently in advance to take protective measures like for instance : stopping placement of quarry run of the core and protecting it with rocks or units according to the wave conditions.

7. Thus installation of the most reliable possible weather forecast system and assistance of a specialized weather organization is essential. Using statistics and data which come to it this organization, during construction, can make short term and medium term weather forecasts : for example 24, 12, 6 hours. The job of the specialist will be facilitated by setting up on the site devices which can transmit data either continuously or at regular intervals : buoys which transmit wave conditions, wind gauge, etc... when the wave generation area is remote from the site, permanent liaison with meteorological centers is required.

Modelling construction procedures

8. The construction methods broadly defined during the design have to be adapted constantly. As a matter of fact during the whole construction period, frequently the quarry is likely to change with the progress of the operation. Moreover the wave climate is not identical all along the structure. Since the construction time can be of several months and even several seasons, it is necessary to take into account the changes of wave conditions with respect to height as well as to duration of storms.

9. The general principle is to know, with the core material, the maximum acceptable wave height for progress of the work in front of the breakwater without important losses and with stable slopes. The effect of tide and the time factor, i.e. the output of the work in progress, must also be considered.

Breakwaters—design and construction. Thomas Telford Ltd, London, 1984

143

10.   In fact there is a whole set of procedures to set up taking into account the following parameters :

- waves (height, duration, direction)
- location and depth of water for the various profiles of the breakwater
- tide
- storms
- available materials
- construction time and organization of work.

11.   Considering the various cases, a study of the construction on a scale model is very useful. This study, using a wave simulation tank, will reproduce the structure in its various phases during construction. Thus, for each part of the work exposed to wave action for significant periods, the amplitude of the waves which it will be able to withstand will be indicated for different directions and for various tidal conditions. The determination of amplitude thresholds depending on the timing and the class of materials considered, even if it does not provide hard-and-fast construction rules, is an element which will be helpful for the site organization to know : use of various materials, storages of various classes, etc... This study will have to be carried out simulating as closely as possible the actual site conditions (materials, equipment transport capacity...) and will also help to produce the work schedules.

AVAILABILITY OF MATERIALS FROM THE QUARRY

Quantity problems
12.   The second problem to be faced by the contractor is materials. A large rubble mound break water requires several millions of tons of materials of various categories defined during the design. If the quarries are not selected and thoroughly investigated during the design process, major modifications of the design may be required during the construction.

13.   Geological and geotechnical surveys are performed for identification of materials. These surveys should examine the extent of exploitable layers and determination of their thickness and stratigraphy. The opening of an experimental quarry remains the best way of defining the operating conditions and of determining the possible outputs and the gradation of materials available from the operation.

14.   The optimum operation of a quarry is that which provides, for each blast or set of blasts, the accurate proportion of quarrystones and riprap which is needed as the construction progresses. On the other hand, if it is impossible for the quarry to yield enough quarrystone (armour) in comparaison with quarryrun this leads :

- either to excess of unsuitable materials and increased expenses
- or to use of concrete units which in each case must be analyzed as to their production and their placing so that the expenses are not increased excessively
- or with similar inconvenience to

surveying remote sites for the required materials.

Thus, it is financially and technically essential to use all that can be obtained from efficient working of the quarries, without excessive sorting of materials or superfluous storage of unsuitable materials.

Quality problems
15.   Nothing is more diversified than rocks and their nature has a great influence on the design. It is important that the quarrystones should not deteriorate under wave and water action and in this regard strength is generally a good indication of durability.

16.   There are a number of tests and standards which may be used to determine the suitability of quarrystone for breakwater construction. (see the paper "Durability of rock in breakwaters").

Transport problems

17.   Special attention should be paid to transport problems. The transport chain is to be defined since it has a large effect not only on the cost of the works, but also in the planning, the construction progress rate and the selection of the installations and equipment to be used at each link of the chain.

WORKING OF A LARGE VOLUME OF CONCRETE

Quantity problems
18.   Very often a rubble mound breakwater is a large scale concrete work involving several hundred thousands cubic meters of concrete (840000 m^3 at Jorf Lasfar, 1500000 m^3 at Antifer, both structures have required artificial rocks). Moreover the cost of concrete works may be more than 50% of the total cost of the structure.

19.   The equipment must be planned to cast several thousand of cubic meters of concrete per day, taking into account hardening time and hazards of construction due to weather conditions.

20.   Several hundred tons of cement are also required every day and this may cause serious problems in some countries and is always an important factor to be considered in construction planning.

Quality problems
21.   Various types of concrete are used for breakwater construction, depending on the nature of work :

- rectangular or cubic units
- multilegged units
- capping, parapet wall
- artificial rock (if any).

22.   The marine environment is essentialy hostile to concrete and provides a severe test of its durability. Thus special precautions have to be taken. (see paper "Concrete and other manufactured materials").

AVAILABILITY OF EQUIPMENT

Construction equipment

23. It is fairly rare that the equipment available has a major influence on the choice of the structure in the preliminary planning stage. The designer first of all tries to optimize the structure based upon the natural conditions, the loads on the structure and the available materials without concerning himself with the equipment which the contractors have or not equipment which it is always technically possible to bring to the site.

24. However, the cost of bringing in the equipment becomes important with heavy or special construction equipment and the size of the structures to be built should be taken into consideration before adopting a design which requires very specialized equipment. For a major project, it is generally advantageous to select a plan whose execution is as rapid as possible with large capacity construction equipment. For a small project, the plan should be implemented with conventional land-based equipment and/or floating equipment which is available nearby. In all cases, when selecting the type of structure, one should take into account the availability of the equipment required to ensure its later maintenance.

Positioning and control equipment

25. For the structures on the site to comply with the project working drawings, it is necessary also to have equipment for positioning and measurements of depths.

26. Electronic instruments are used more and more to replace other alignment methods for setting out. According to the circumstances, an existing infrastructure is used or a short range portable device is installed. The accuracy of positioning obtained with these methods depends upon the location of land-based stations and, when they are carefully selected, accuracy can be of a high order. The LASER should also be mentioned although its beams are difficult to use on floating apparatus. It is a valuable aid for alignment and altitude adjustment.

CONSTUCTION BY FLOATING EQUIPMENT

27. Effective construction at sea is possible only if the wave climate is not more severe than it permits operation of floating equipment a large number of days on year round basis or a seasonal operation with practically no interruptions.

28. Practical execution of most marine construction is possible only if one has the use of land facilities, a service port and powerful floating equipment capable of dealing with high peak workloads. This type of equipment may involve large costs, it is therefore mainly useful for large construction works, not for smaller structures.

29. Very often the most economical method is construction from the land but it is not always possible to do so and both approaches from the land and from the sea are required. Use of floating equipment is contemplated for :

- preparation of foundations
- dredging
- dumping
- placing the armour units

Preparation of foundations

30. If the mechanical properties of the subsoil are such that penetration, slope failure or settlement can occur, special preparation of foundations is required.

31. A first method is to place a bottom mattress and to wait for the consolidation of the soil before building the upper part. This method is possible only if the construction schedule allows for and if the mattress is not exposed to deterioration by wave action during the pre-consolidation process.

32. When it is necessary, unsuitable soft material can be dredged out and replaced with better quality material such as sand, gravel or a mixture of both. This method may be very costly.

33. There is in addition a range of expedients which may be adopted for increasing the shear strength of weak soils in situ including the following :

- Driving a large number of small piles at close centres
- Dynamic consolidation by means of the impact produced by heavy weights on the bottom
- Placing column or piles filled with cohesionless materials such as sand or gravel in order to accelerate drainage and consolidation of soil in situ.

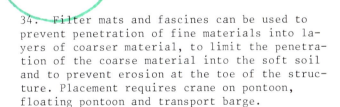

34. Filter mats and fascines can be used to prevent penetration of fine materials into layers of coarser material, to limit the penetration of the coarse material into the soft soil and to prevent erosion at the toe of the structure. Placement requires crane on pontoon, floating pontoon and transport barge.

35. All these methods can be difficult to perform in deep water with floating equipment working when the sea condition is good.

Dredging

36. Dredging equipment can be used for preparation of the foundation by dredging out the weak soil and for placing sand and gravel extracted from offshore resources.

Dumping

37. As a general rule, most of the underwater part of the rubble mound structure can be built up by dumping from various kind of barges. This method is economical for large quantities of materials.

38. The advantages of placing quarrystones by

dumping from floating equipment include the following :

- dumping with modern floating equipment can be achieved with higher wave heights than with land-based equipment.

- floating equipment is stronger than land-based equipment.

- construction of the most exposed upper port from land is easier and faster when a large part of the underwater portion is built from the sea.

- the lag between execution of the underwater part from the sea and construction of the upper part from land may reduce further settlement.

39. The drawbacks are the need of a service port for shelter and loading of the materials and the inacurracies of implementation and monitoring.

40. Dumping equipment
Carrying and placing quarrystone from the sea is usually done by :

- hopper barges (or hoppers)

- dump barges

- ordinary barges

- push-off barges

- "hydroclap" barges with full opening (split-hull barge).

41. Hopper barges carry the materials in compartments that can be emptied underwater by opening the bottom of the compartment directly over the desired place. These hoppers may be self-propelled or towed : the largest can release several thousand tons of materials at a time.

42. Dump barges carry the materials on deck and have a ballast system (controlled by valves) that enables them to roll sharply and dump their loads over the side. The largest can dump several hundred tons of materials at a time.

43. Ordinary barges differ from dump barges only in the absence of controlled tipping. Some are however self-propelled. They are unloaded by crane, loader, bulldozer or by hand. In this case placing takes time.

44. Push-off barges dump the materials from the deck by pushing, on either side of the axis of the barge, with a bulldozer type blade. They can lay materials in thin and homogeneous layers by successive passes using a simultaneous control of propulsion speed and of dumping speed.

45. Hydroclap barges are made up of two watertight caissons to hold the materials, connected by a hinge, with an axis parallel to the axis of the barge, located at the deck level. This device makes it possible to totally open the materials compartment and the draft of barges is minimized. The push-off barges and hydroclap barges have bow and stern propellers steerable

in every direction, which provide excellent manoeuvring capabilities. Hopper and hydroclaps give the cheapest cost per ton of materials set in place.

46. The loading system also affects the cost and it should be carefully optimized. It can be as follows :

- by direct dumping from trucks,

- by loading with a crane, using for example, buckets with the same capacity as the truck,
- by conveyors

47. The choice depends upon the volume and size of material. In the first case, the barges must have an appropriate structure so as not to be damaged during loading. The tides must be taken into account because they can greatly affect the drop height of the material. The second case is applicable when very high daily output is required and when the overall quantities are very substantial, so as to reduce the impact of cranes in the overall cost.

48. Accurate placing of quarry materials by barges depends on :

- the accuracy of positioning equipment which relies on the methods for relative location (optical procedures, radio locating), on the site conditions (distance from reference marks or transmitters, currents,...) on the manoeuvrability of the barges and on the speed of dumping.

- the particle size, the finest materials may undergo excessive dispersion due to current action, particularly in deep waters.

49. Besides, it is advisable to allow for inaccuracy in the surveys done by sounding equipment, themselves subject to positioning difficulties. All these elements put together imply that it is wise, for some parts of the structure, such as substructure and toe abutments, to provide supplementary width beyond what is strictly required. Moreover, it is often not easy to carry out as many controls as necessary either because of the construction procedure itself which imposes the almost immediate covering of vulnerable materials with layers of large particle size, or because of the sea conditions. So, it is difficult to estimate the quality of construction of a structure and it is necessary to advance while complying with the quantity criterion.

50. For this purpose, whenever possible, determine the apparent density using full scale tests on land for each class of materials. Next, for a specified length of the structure place a fixed quantity of materials which will be at least equal to that which the geometry of the work imposes, but will not exceed that theoretical quantity by more than 10 to 20% compatible with the part of the structure concerned and with that of adjacent sections. To this is later

added, in the long term, an internal settlement of the mass (consolidation) on the order of 1 to 2% (or even 5%) of the structure height.

51. Core fill can only be placed up to a certain elevation e.g. about 2 meters below water table by split-hull barges. Other barges may allow higher placements depending upon the wave action.

Placing the armour units

52. For heavy quarrystones and concrete units, direct dumping from barge is not possible. Placement may then be from floating cranesprovided the weather is calm. This may be the best solution if a gentle slope is required. Nevertheless this method suffers hazards of positioning close to the structure and of breaking the units during placing under wave action.

53. Use of self-elevating platforms which can move in steps of 4 m is very practical for all kind of breakwater work. The platforms, however, have to be supplied with materials by barges.

CONSTRUCTION BY LAND-BASED EQUIPMENT

54. Construction of coastal structures from land utilizes general construction equipment that is cheaper than the equipment required for water-borne construction. It is also, easier to re-use this equipment for other inland construction. But it would be a mistake to believe that the building of harbour and coastal structures from land is completely without hazard.

55. Construction is dependent on the sea conditions since the personnel at work on the structures may be endangered by heavy weather, and since extremely rough water may dislocate all or a portion of the materials recently placed in the water from land before it has been protected. On the other hand shore-based cranes cannot armour sea-placed core stone until it is completed to full height and isolated dumps of core material become increasingly vulnerable to damage from tide rips through gaps which are being progressively closed. Moreover, it is to reduce these risks as much as possible that it is important, as previously emphasized, that a good weather coverage is available. On the other hand too pessimistic weather forecasts lead to additional expense involved by a stoppage of the work and to place a superfluous protection, with heavier materials.

56. Work done solely from land must be done progressively whereas when the work is done by sea, it is possible to work in parallel, with several floating items of equipment in use at a time. This explains why it is impossible to lift very large loads and to carry even small loads very long distances when the work is done from land. There are also some difficulties in the use of long-wheelbase rail cranes and turning around trucks and other machines that must often cover distances too great for reversing in order to return to land. This means that it may become necessary to make the structures wider than they would have to be to withstand wave action.

57. In this respect, in order to cope partially with this problem the use of turning platforms can make it possible for trucks to turn back on breakwaters -in restricted areas- and thus make manoeuvring easier. Shifting the platforms as work progresses does not raise any special difficulty since the equipment used to do so such as cranes or graders is normal plant on breakwater construction. The time needed for that operation usually does not exceed two hours. The optimal distance between the platforms and the heading face ranges between 150 and 300 metres. Thus the section to be covered backwards by lorries is reduced to a minimum and at the same time the work, carried out at the breakwater heading face proceed normally.

58. Usually, the traffic of haulage units and the movements of earthmoving and handling equipment must be performed at an elevation high enough above wave action so that risk of overtopping does not impede the construction. One method is to raise the level of the core crest above the highest water level before building the superstructure. On the other hand if the armouring has kept pace with the core construction as recommended it can provide the necessary protection to a roadway at a lower level. If the constraints affect the width and level of the crest of the breakwater, they will result in a substantial increase of the volume of materials to be placed. However, it should be noted that this increase mainly affects the core materials, that is the less expensive and usually most abondant materials.

59. There is also an imperious requirement in order to make sure that construction from land is performed under proper conditions : every effort must be made to keep traffic lanes in good condition whatever they are (quarry lanes, breakwater access, graded traffic area on the structure). If it is very rainy on the site and if there is fall off from the vehicles which might form mud, it is absolutely necessary to provide cleaning equipment (bulldozers and graders) and to bring in materials regularly to preserve good road surfaces. This maintenance aspect is more important than it may seem in the organization of a major work site, since it can greatly affect progress and it is rather expensive.

60. It seems necessary to distinguish several case within the framework of building sloping protective structures from land according to whether the sea climate hampers the work by sea during long periods of the year for the deep portions of the breakwater (particularly for foundation mass).

Sea conditions preventing the use of floating equipment during long periods of the year :

61. In this case, construction by sea is generally limited, during good weather, to building of breakwater foundation with hopper barges up to the maximum possible level or a lower level determined by the stability of the material to resist damage resulting from the periods of bad

weather. The upper portion of the structure and the root of breakwater which preclude the use of water-borne equipment are to be carried out from land.

62. If the sea conditions are rough (practically constant swell and significant height HS higher than 2 m, for example), it will be absolutely necessary to place all the materials with cradle or grab, with a high capacity crane running on a concrete cap built gradually as the work progresses. Hight capacity crawler cranes are available which are very appropriate, since they can move quickly along the breakwater and can be easily pulled back when a storm is forecast.

63. The core material should be covered as soon as possible with a layer of quarrystones forming filter, then by quarrystone located immediately under the armour and, finally, by the armour units. The partial construction of the concrete cap makes it possible to protect the crest of the breakwater which is very vulnerable during storms.

64. Usually and for difficult working conditions, materials with the highest possible specific gravity and rocks with high unit weight are used. Since the distance to quarries for obtaining these materials may lead to considerable construction costs, every effort should be made to achieve the minimum profile able to withstand the action of design storm and still allow for movement of construction equipment.

Moderate sea conditions during long periods
65. Two assumptions should be considered depending on the cost of quarrystone materials, and on the distance to the quarries.

Nearby quarries and inexpensive quarrystone materials
66. The structure should have a crest wide enough for easy movement of haulage units. The material for the core, the underlayers, and the armour underlayer are placed by direct dumping from trucks and pushing with bulldozer. The slope under the armour or that of the underlayers may be completed by crane to obtain the planned gradient. The lower part of the structure (unless they have already been achieved using hopper barges) and the armour are placed either with a rail mounted crane, or heavy crawler cranes. The armour should be placed as close as possible to the advancing leading edge of the structure so as to minimize exposure of the materials to damage by storm wave action.

67. In some cases, it may prove necessary to build a lean concrete working surface on the crest which will improve the stability of the crest and provide a suitable roadway. In any case, during moderate weather, one must work around the clock (with perhaps a shutdown of a few hours for maintenance of equipment). In areas characterized by a high tidal range, this may be taken advantage of, to spread out the materials and to obtain a more gentle slope at the structure head.

Distant quarries and expensive quarrystone materials
68. In that case, the solution which may prove imperative will consist of building a minimum profile and to place all the materials (quarry-run , rocks and armour units) with the placing equipment (rail mounted crane or heavy crawler crane). To reduce the working period, and under assumed relatively favourable sea conditions, it would be generally advantageous to build the largest portions of foundation mass with hopper barges.

69. The upper portion of a sloping breakwater can be constructed more accurately when built from land because it is possible to monitor the construction tolerances and placing techniques. The permissible allowances for the exposed portions should not be greater than those which are usually taken for earthwork. It is somewhat difficult to place the riprap layers on the slope It is necessary to know the actual gradient of the core slope which determines the quantity of riprap to be placed layer after layer. There again, the criterion of quantity noted for the placing of materials by sea should be used, the possibility of examining the emerged portions also making it possible to better estimate the distribution of the riprap on the slope. Section of some structure parts can be surveyed by means of land-based cranes used for their construction. Thus the slope and the thickness of the various layers can be checked.

70. First of all, it is advisable to perform full-scale tests of placing riprap on a slope under conditions such that the results are measurable. When necessary laboratory tests on full-scale model may result in better understanding of the way the rocks take up their position and particularly, of the effects of water damping phenomena which might adversely affect the homogeneity of the thickness of the layers. These tests will make it possible to optimize the particle size of the riprap and to determine the sensitive points where grading is to be performed.

71. Thus, the allowances on slopes will be defined from information resulting from the above-mentioned tests or, as in many cases, from information acquired during previous works where the same materials were used with the same particle sizes. Allowances will be set in volume, or, more generally, in weight, in comparison with theoretical volumes or weights, and allowing for the gradients likely to be taken by slopes built under normal conditions. These normal conditions mean a structure built applying required precautions : a given class of materials should not be subjected to wave action beyond the height determined by scale model tests above which deterioration of the masses would impede achievement of satisfactory geometry.

Placing the units
72. There are various ways of hooking concrete units to the load beams of sheer hulks (steel staples, lewis holes, or grooves for cables or chains), but it is important for them to allow the unit to be unhooked and rehooked in a wide

range of positions, both to be able to correct excessive placement errors during construction and for subsequent maintenance, repairs or reinforcement of the structure (steel staples are best in this respect, although they have a limit of about 40 T per fastener). For handling squared blocks, intended to be arranged randomly, it is also common to use self-tightening tongs which have the advantage of considerable speed in handling. However, this equipment is not designed for rehandling the units. For lifting special shaped concrete units like tetrapods, dolosse, etc... there are special hooking systems, suited to the design. For instance, tetrapods are hooked with a single sling passed around a lug ; on the other hand, for the dolos, a special grab bucket is used which allows, among other features very fast unhooking of units.

73. Placing is generally carried out by cranes on rails or by large crawler cranes. One of the characteristics of such equipment should be quick pullback in case of storm, in contrast to the use of rail-mounted cranes which have a major problem since the rails themselves may be damaged during storms. Modern crawler type cranes with ringer and skyhorse have a momentum capacity up to 3000 TM.
On structures whose dimensions are not very large backhoe excavators are also used for placing rocks and grading of underlayers. Specialized machines exist which can lay natural quarrystones of 15 T.

74. The placement procedure must be chosen in order to achieve the specified number of units per unit area and interlocking. The required use of placement schemes results from scale models experiments run in hydraulic laboratories.

Equipment for extraction and land transport
75. Quarry materials are extracted and transported with conventional civil engineering equipment. The present trend is to use large capacity dumptrucks (25 to 50 T). A large transport capacity is helpful for intensifying protection measures quickly in case of storm forecasts and also to accelerate construction when conditions are favourable so as to reduce the time lost due to bad weather.

CONCLUSION
76. Design and construction cannot be separated. During the design, construction and maintenance procedures should be taken into account in order to design a structure which can be executed and maintained effectively as closely as possible with design.

77. During the construction close cooperation between contractor and designer is vital in order to adapt the construction work to the actual weather conditions and operation of the quarry without endangering long term stability.

78. During construction and after completion of the construction work continuous recording of physical data and monitoring of all parts of the structure are required.

REFERENCES

1.PIANC Final Report of the 3rd International Commission for the study of Waves. Bulletin of PIANC n° 36 (Vol II/1980) Supplement 1.50.

12 Rubble breakwaters—specifications

P. G. R. BARLOW, MA, FICE, MConsE and M. G. BRIGGS, DLC, FICE, MConsE, Coode & Partners

---

This paper deals with the specification of materials and workmanship in the construction of rubble mound breakwaters with particular emphasis on such factors as size of stones, control of placing of stone and allowance for settlement. No universally accepted standards are in existence for some of the items covered by the paper and it is recognised that some of the suggestions are controversial. It is hoped that the paper will provoke discussion, leading to a consensus view on an improved specification.

1. INTRODUCTION

1.1 To the uninitiated a rubble mound breakwater is an unsophisticated structure, apparently a heap of stones put together haphazardly and not to be compared with the solid gravity wall structures protecting our older harbours and familiar to the general public as engineering achievements in man's battle against the elements.

1.2 In fact the rubble breakwater, properly designed and constructed, is quite a sophisticated structure and represents one of man's better achievements in controlling the elements. The emphasis, however, must be upon proper design and construction, and when these are achieved, together with a proper assessment of the sea conditions to which the structure must stand up, the rubble mound breakwater performs its function admirably and has the added important advantage over other more solid structures that such failures as do occur are seldom total failures and the remains of the structure can normally be built up again to perform their full function without undue difficulty. The main reason for this is that an under-designed breakwater of this type can be rectified by flattening the slope on the seaward side, and in the course of failure, the sea usually performs this function, so when the top part is built up again, the structure is even more resistant to wave attack than formerly.

1.3 Advantage was taken of this particular attribute of rubble mound breakwaters in the early days of such works, when the process of trial and error was an accepted method of solving sea defence problems and when large works were attempted and achieved with minimal plant and sophistication.

1.4 In a report of 1880 on a harbour in New Zealand which was to be built by direct labour, Sir John Coode wrote:-
"Whilst, however, expressing the fullest confidence in the permanent stability of the work after the slopes have been flattened down to the inclinations shown, I desire to point out that during execution some fears will almost inevitably be engendered as to the sufficiency and ultimate permanence of the structure, in consequence of the occasional and recurring flattening down of the slopes by heavy gales during the process of forming the mound. But it must be borne in mind that the agency by which the material is distributed and trimmed to a proper slope is one of wave-action, the operations of the workmen being confined to depositing the stone, so that it shall ultimately produce a mound when "clawed" down by the sea with the least possible waste of material – a matter requiring care and judgement where the quantities to be dealt with are so vast. However much the apparent dislocation of the mound during progress may appear from the surface, it may be taken for granted that the action of the sea will only tend to distribute the rubble over the area required for the base of the work, and that the mole when finished will, as I have explained, partake very closely, if not actually, of the profile shown on the cross-section although the seas of several winters would be necessary for the production of the ultimate slope as shown. In the meantime the rubble would be tipped on the top of the bank for the subsequent "feeding" of the slopes by the sea".

1.5 Nowadays, the lesson that has to be learned from this is that the best economy is achieved if the Engineer is not too rigid in his theoretical approach to the problem of breakwater design and is prepared to adapt to natural local conditions. In particular he should adjust his specification to work with the sea rather than against it and use the best materials which are locally and economically available rather than insist upon achieving an unnecessary high theoretical quality of the work at greatly increased expense.

2. SPECIFICATION

2.1 General

2.1.1 In present times most large works are

Breakwaters—design and construction. Thomas Telford Ltd, London, 1984

151

constructed by contract. There is therefore less room for ad hoc changes during the construction period than would be acceptable with the direct labour system. The Engineer is faced with the problem of giving the Contractor who is tendering for the project a clear statement of what he has to provide, while at the same time making it clear to him that sufficient flexibility will be allowed in the Specification to cater for changes that are revealed during the Contract period and for making the best and most economical use of the materials that are actually obtainable within reasonable distance of the site.

2.1.2 Certain elements of the Specification are quite straightforward. The drawings will show the line and profile of the breakwater and will detail the filters, core and armouring layers. The changes in section, which the Engineer requires to cater for changes in depth or in exposure, will also be shown, as will details of any wave walls or other structures on the breakwater. Thereafter, there are three main items which are much more difficult to specify satisfactorily. They are the size and quality of material, the method of placing it, and the degree of variation acceptable to both. It is on these three important points that this paper will mainly concentrate.

2.2   The Basic Design
2.2.1 The design criteria for rubble mound breakwaters are well enough known not to need elaboration in this paper and have, in fact been covered in other papers. Suffice it to say that for a given density of armour stone the two most important elements in the design of the exposed face of the breakwater are the weight of the armouring stones and the slope to which they are laid. From this it follows that the optimum design is achieved when the right balance is struck between the total quantities of rubble (which increase as the slope is flattened) and the maximum size of armour stone that can be economically obtained in sufficient quantities in the quarry.

2.2.2 Beneath the main armouring of the breakwater are various other layers of material and ideally these should be selected so that the best possible use is made of the materials economically arising from the quarrying operation.

2.2.3 The smallest stone, which arises in the largest quantity is used for the core or hearting of the breakwater and this is protected by several layers of gradually increasing size. The essential elements in all this are that each layer shall be of such a size that it interlocks effectively with the layer below it, and that the relative sizes between layers shall be such as to prevent the stones from the inner layers being drawn through the outer layers. Additionally, it is essential that the outer layers of the breakwater should be so arranged that a satisfactory void ratio is maintained, because it is this labyrinth of voids which contributes most effectively to the absorption

of wave energy.

2.2.4 The specification of the various items in the breakwater is to a large extent governed by the basic design essentials outlined above.

2.3   Selection of Rock
2.3.1 Ideally, the type of rock that will be used in the structure should be known before the design is prepared and the actual quarry sites should have been thoroughly explored and tested before the project is put out to tender. Unfortunately, this is not always possible to achieve, for a variety of reasons, such as shortage of time, unavailability of funds, and the fact that free access to the quarry site is often difficult to arrange except in the role of a potential purchaser of the land. Furthermore, it is often not possible to assess the productivity and the relative sizes of stone obtainable from a given quarry until it has been properly opened up. Thus, the specification must be somewhat general in character and although the Engineer must be sure of the general quality of rock that is available before he designs the structure and specifies the required rock, the choice of quarry normally has to be left to the contractor.

2.3.2 The basic specification for the rock must be aimed at ensuring that the material will be sound enough to be placed into position without fracturing, and durable enough to resist weathering and abrasion for its economic life in the breakwater. Hardness, toughness, and soundness are the three qualities that must be secured, and there are tests specifically aimed at proving these characteristics. In specifying these tests, it is most important that care is taken to ensure that samples are representative, and whatever tests are specified it is important that practical experience is used to check the results. One of the most important tests that can be applied is to examine the performance of the selected rock if it has been used in other works. The geological examination will typecast the rock and will indicate what further tests are required. A full petrographic examination is recommended and this will often indicate the amount of weathering which may be expected. The degree of testing depends very much upon the local circumstances. If, for example, granite is readily available, the examination probably needs to be a good deal less rigorous than in the case of, for example, limestone. Specific gravity and water absorption give a good indication of the hardness and soundness of the rock and obviously where absorption is high and specific gravity is low there is a clear indication that the stone is of marginal quality and requires great care in supervision and testing.

2.3.3 When particular figures are entered into a specification it is most important that the Engineer should be satisfied that they can be met, and in the following example which the authors have recently used this was the case:-
All rock for the breakwaters and slope protection shall be hard dense, non-friable stone

from sources approved by the Engineer. The rock shall be free from cracks, seams and similar defects and shall not fracture when dropped through 1.5 metres onto a steel plate. Rocks shall be of all sizes within the limits specified and shall be angular in shape, being neither unduly elongated or flat, nor unduly rounded. The greatest dimensions shall be less than twice the least dimension. Rocks which do not comply with these requirements will not be accepted in the permanent works. The specific gravity of the rock on an oven-dried basis, used in the breakwaters and slope protection shall be not less than 2.65 and the water absorption not more than 2.0% of the dry weight.

2.3.4 One general point that should be made in regard to the specification of stone for the breakwater, which applies particularly to the armouring, is that the specific gravity of the material needs careful attention. The reason for this is that a relatively small variation in specific gravity has a relatively large effect on the weight of armour stone required. In a typical case, for example, the change in specific gravity from 2.7 to 2.5, (a reduction of 7.4%) resulted in the nominal armour stone size having to be increased from 8 tons to 11.5 tons (an increase of 44%). The reason for this large increase is that the resistance of the armour to wave attack is a function not only of its weight but also of the surface area exposed to wave attack.

2.3.5 The importance of this point is that if the specific gravity is stipulated at too high a level by the designer, and it is later found impossible to achieve this value, the whole geometry of the breakwater profile is affected because the change in stone weight also affects the thickness of the armouring layer. It is therefore prudent to err on the side of under-estimating rather than over-estimating the specific gravity.

2.3.6 The shape stipulation of 1:2 for least to maximum dimension could be varied in situations where it was not practical to achieve the figures stipulated, but we would not normally recommend anything in excess of 1:3 because this tends to produce instability or undue irregularity in the final breakwater cross section. The water absorption and specific gravity were also fixed in this particular case to suit the local conditions, but these will vary very much from site to site and the specification should in fact be directed towards pointing the contractor in the direction of the most desirable, not necessarily the cheapest, stone available.

2.3.7 It is worth noting that the outer layers of the breakwater require somewhat more stringent specification than core and inner layers and it would be wrong to apply too rigorously the standards of the outer layers to those of the core if by so doing the use of materials arising in the quarry was thereby prevented.

2.3.8 A valuable paper given by P.G.Fookes and A.B. Poole on the suitability of rock for breakwaters has been published in the quarterly journal of the Geological Society of London (1981), Vol. 14 pages 99-128, which is recommended as a guide to the quality of rocks found in the marine environment.

2.4 <u>Sizes of Rock</u>
2.4.1 In the experience of the authors it is in the proper construction of the core of the breakwater that most of the site arguments between contractor and engineer occur.

2.4.2 It is arguable whether the core should be well enough graded to make it virtually impermeable or whether it should be deliberately constructed with voids as in the case of the outer layers. In fact circumstances may decide which of the two is appropriate (e.g. whether or not the breakwater is immediately adjacent to reclamation or whether the transmission of any wave energy through it is acceptable). In either case, however, it is considered quite unacceptable to have large quantities of small or friable material tipped haphazardly into the core in such a way that it can form large concentrations of small material which can eventually wash into the voids causing undue settlement, possibly after the construction of the breakwater is complete. For this reason it is considered by the authors to be essential that the contractor installs a proper system for grading and selecting the core material. In our experience, most of the problems on site occur when the contractor tries to select the material from mixed stockpiles merely with the aid of excavating plant. In our view some form of screen is essential.

2.4.3 Having said this, however, it is emphasised that proper consideration should be given to the Contractors' problem in making economical use of most of the material which arises during the quarry operation. For this reason, we recommend allowing the use of smaller stone sizes in the inner part of the core than in the outer core layer.

2.4.4 In deciding upon the maximum size of stone permissible in the core, regard must be had to the degree of exposure of the site. In a relatively calm situation we would regard about 1 tonne as a suitable maximum size, but in the case of a site where rough weather can be expected before armouring is complete we would suggest going up to about 2 tonnes. This would give the Contractor a chance to keep the hearting reasonably stable before the armouring layers are placed upon it. In other words we suggest that the upper limit should be governed by site conditions and that it is not particularly critical provided that the outer surface of the core material is reasonably uniform in roughness. It is suggested that the general specification of the core material should be from about 10 kilogrammes to the upper limit of 1 or 2 tonnes, reasonably uniformly graded and that this specification should be rigorously applied to the outer layer which we suggest should vary from 1 to 2 metres in thickness, depending upon the degree of exposure of the site. For the central core of the breakwater, that is inside the outer layers described above, we suggest that the minimum size of stone acceptable could be reduced to as little as 1 kilogramme but not less, as stones of less than 1 kilogramme in weight, in any significant quantity, could present a settlement problem.

2.4.5 It must be borne in mind that the above suggestions for specification do depend upon the breakwater being of reasonably large general cross-section. Obviously if this is a very much smaller breakwater the stone sizes must be adjusted to suit. The main principle to observe and to be covered in the specification is that the stones should be large enough to remain in position, well enough graded to inhibit undue settlement and that the outer face should be composed of stones large enough to provide a rough surface for the next layer above.

2.4.6 We do not propose to specify at this stage the thickness or thicknesses of the intervening layers. These in fact will be governed by the availability of stone, because as has been stated previously their main purpose is to provide a well voided transition between the core material and the heavy stone in the primary armouring layers, making best use of the stone sizes arising economically in the quarry.

2.4.7 For the primary armouring of the breakwater we would look for the largest stones that can be economically quarried in sufficient quantity to provide a uniform covering.

2.4.8 As for the thickness of the main armouring layer, we suggest that it should be such as to accommodate at least two stones of the specified size, nested together in a practical manner. Clearly this is something less than twice the maximum overall thickness for each stone.

2.4.9 Although the main armouring is considered as a single size stone, it is usual to specify the upper and lower limits, and the normal tolerance is approximately ±25% on the nominal weight. Thus a nominal 8 ton stone would be specified as 6 - 10 tons. It is considered necessary to go one step further in this specification and guard against the possibility that most of the stones will be at the lower end of the range. This can be achieved by specifying that a certain percentage of the stones are to be equal to or greater than the nominal size, and for the outer layer, assuming that the nominal size is the size required to satisfy the design formula, we suggest that a figure of 75% or thereabouts would be appropriate.

2.4.10 From the design point of view, the size of stone in the layer beneath the primary armouring should not be less than 1/10th in nominal size in the latter. However, to make the most effective use of material arising in the quarry it is often desirable to specify much larger stones than this in the secondary layer. In any case, we would suggest specifying the nominal size ±25% or 30%

2.4.11 With regard to the foundation of the breakwater, the ground conditions will determine whether the core will be placed directly on the sea bed or whether a filter is required. The object of the filter is to prevent the core from settling into the bed material and the grading for the filter depends on the sieve sizes of the two materials. If the bed material is very fine, it may be necessary to form a double layer filter. The material for this filter is screened out at the quarry and the grading can be laid down in the specification in the same way as that for concrete aggregates. The filter is laid as a blanket over the seabed and we recommend that each layer should not be less than 300 mm thick.

2.4.12 As an alternative to the stone filter, fabric filters are often used. The specification for these again depends on the composition of the seabed which determines the size of mesh. Its strength properties must make due allowance for the method of handling and the ability to resist puncture, and consideration must also be given to the properties of the material of the fabric which must be durable to ensure permanence of the structure. It is a wise precaution to add a layer of protective stone on top of this fabric filter and this too should be no less than 300 mm thick We suggest that it could consist of material up to about 75 mm in size, the purpose of the layer being to prevent rupture of the fabric when the core stone is tipped into position.

2.5 Placing Stone

2.5.1 Normally the method of placing the core material is left to the discretion of the contractor, with few restrictions, provided he uses material of the specified sizes and produces the correct profile. The designer should give careful consideration to the contractor's probable construction methods, particularly in respect of the level and width of the top of the core berm and the slope of the sides. For larger structures it is common for the contractor to construct the lower part of the core by means of bottom-dumping barges, but the final trimming and all of the upper part is normally formed by tipping from end or side tipping lorries (preferably both) aided by excavating plant for final trimming.

2.5.2 It is not normally cost effective to design the core wide enough to permit lorries to be backed and turned, so the contractor is usually permitted to incorporate occasional widened sections as turning bays, and these usually have to be trimmed back to the general profile before armouring is placed

2.5.3 It is suggested that the method of placing stones in the intermediate layers depends upon the maximum size, and that tipping or random dumping should be permitted up to a maximum stone size of about 2 tonnes. For larger stones, and certainly for all the exposed stones in the outer layers, these must be placed individually. When large stones are dumped indiscriminately in large quantities it is inevitable that bridging and arching will take place and that substantial settlement will occur in due course when the section is exposed to rough seas.

2.5.4 The correct placing of armouring stone involves a considerable measure of skill, the intention being to produce a generally neat profile composed of well interlocked stones and presenting a rough surface to the sea.

2.5.5 The authors suggest that the placing of stones should be covered by general clauses such as the following:
"The effectiveness of the breakwater is dependent on its void content and permeability, and the sizes of the rocks in each part of the breakwater are essential features of the design. The stability of the rubble slopes and their resistance to attack by the sea are dependent on the effective interlocking of one stone with another and of each layer with the layer beneath. In addition, the surface layer, though neatly laid to the approximate profiles required, must present a rough, uneven face to the sea, avoiding any concentration of flat surfaces so that the waves are absorbed by the breakwater instead of being deflected over it"

and

"The methods used for placing the rock in the breakwaters shall be subject to the approval of the Engineer. Rock placed directly on filter material shall be placed in such a manner that the filter is not ruptured. Apart from that, no particular restriction will be applied to the method of placing the rock in the core or in those inner layers of the breakwaters of which the maximum stone size is less than 2 tonnes, the rocks for which may be placed by tipping or dumping from barges, once the filter layer has been protected to the satisfaction of the Engineer. All rocks in the outer faces of the breakwaters, and those which are in excess of 2 tonnes in weight in the inner layers shall be placed individually, and tipping or dumping will not be permitted".

2.6   Survey, Inspection and Measurement
2.6.1 The problems of survey, inspection and measurement are inter-dependent, and should be closely considered when the contract documents for any particular site are being prepared. Although, by definition breakwaters are only required in places subject to storm wave attack, there are some sites where the weather between storms is extremely calm for long periods, so that close control of rubble-placing by normal sounding methods is very straightforward. In other places there is a continuous heavy swell throughout the year, augmented from time to time by severe storms, and in such situations the control and checking of work is very difficult.

2.6.2 In conditions where check sounding is reasonably straightforward, and where early displacement or consolidation of the ground is not expected to be a major factor, it is recommended that all billing and measurement should be based on the volume of the theoretical profiles, applied to the actual sea bed contours as revealed by a pre-construction survey. This places the onus for assessing losses due to settlement and consolidation on the Contractor. The responsibility of the Engineer is to ensure that the structure is built to the correct profiles, and it is most important that the work is checked at each stage, and not only after completion. If the core, for example, is constructed to less than full profile the breakwater may be unduly permeable, and if the core is substantially over profile this will result in a corresponding reduction in the volume of the underlayers and armouring, unless special measures are taken to maintain the specified thickness of the outer layers.

2.6.3 Where conditions permit frequent, reasonably accurate, check sounding, it is only necessary to draw attention in the tender documents to the way in which the quantities will be measured and to make it clear that the Contractor will have to provide the necessary instruments and assistance to the Engineer's Representative in order to check each phase of the work.

2.6.4 Setting out, monitoring and measurement usually involve an element of ingenuity in adapting the plant and instruments available to carry out the tasks required in the particular conditions of a given site, and one can only indicate the sort of methods that are commonly used:-

(i)   Sounding chain adapted with large "basket", to bridge gaps when checking armouring layers. (In general, echo sounding is inappropriate except possible in the early stages, for checking filter layers or barge-dumped material, when the sounding launch can pass transversely across the structure).

(ii)   Buoyant level-staffs, with broad foot or "basket".

(iii)   Timber or wire profiles extending some distance below surface and if possible down to the sea bed as a guide to diving inspectors, where visibility is adequate to inspect underwater profile for major irregularities.

(iv)   Horizontal beam carried by rubble-handling crane and used for profile distance measurement.

(v)   3 - 4 m long rod of neutral buoyancy, used by divers to detect large hollows or protrusions when inspecting profiles by feel, in conditions of bad visibility.

2.6.5 In situations where swell conditions are unrelenting enough to prevent effective setting out and checking by methods such as those mentioned above, it is important that this fact be recognised when the tender documents are being prepared. Since it is not normally desirable to restrict the contractor's freedom to determine his own working methods and choice of plant, the tender document should draw attention to the problem of controlling rubble placement, and should require the Contractor to submit with his tender, for the Engineer's approval a clear description of his intended working methods, indicating how he proposes to control the work and what facilities he will make available to the Engineer's Representative for checking and measurement.

2.6.6 In conditions of permanent swell it will be impractical to complete the core and other profiles from floating plant and the contractor will probably resort to the use of a crane mounted on a gantry, moving forward on the completed core, or on a jack-up platform alongside the work. In either case the control of rubble placement can be effected entirely by instruments related to the jib angle and outreach and the length of wire paid out. The same plant can be used to check the completed profiles, with the aid of some form of level-staff suspended from the crane job, but as this will inevitably interrupt the work of placing rubble it is important that the specification ensures that the Contractor allows for this essential checking operation in his programme and rates.

2.6.7 In view of the difficulty of accurately checking the outer armouring profiles by sounding, it is recommended that an additional check is made by controlling the quantity of armour stone placed per unit length of breakwater.

2.7 Settlement

2.7.1 Settlement in a rubble mound breakwater is attributable to four factors:-

(i)   Shaking down of the rubble.
(ii)   Penetration into the sea bed.
(iii) Displacement of soft sea bed material.
(iv)   Consolidation of soils beneath the base of the rubble mound.

2.7.2 The objective of the Specification should be to ensure that the breakwater is complete to the full design profile at the end of the maintenance period. Settlement after the end of the maintenance period, due to the long term consolidation of soils beneath the base of the structure, should be provided for in the determination of the design profile - that is to say the crest level should be set high enough to remain at or above the height required to satisfy hydraulic criteria when long term settlement is complete.

2.7.3 Items (i) to (iii) above require to be provided for during construction, because, unless special measures are taken in anticipation, it is probable that the levels at the end of the maintenance period will be below the design profile, but not sufficiently far below to accommodate another complete layer of outer armour stones. Thus, unless major lifting and

re-laying of stones is to be carried out, the profile will end up either marginally below or will have been made up substantially above profile.

2.7.4 The authors have customarily covered the point by a clause such as the following in the Specification:-

"In placing rock in the breakwaters it is assumed that settlement will in due course cause a reduction in the height of the structure. To compensate for this the Contractor shall place the rock to levels and sideslopes which correspond to vertical thicknesses 10% greater than the designed vertical thicknesses as shown on the drawings. No additional payment will be made in respect of this increment."

2.7.5 It is arguable whether this allowance of 10% is of the right order or not, but it is clear that the final result is very sensitive to the speed of construction and to the degree of exposure to rough weather during the construction period. If, for example, a long length of breakwater core is constructed and covered with the armouring layers, all within a calm weather period, the total settlement that eventually occurs may well reach or even exceed the 10% figure. If, on the other hand, the core is pushed forward against difficulties during a rough weather period, the preliminary shaking-down will occur simultaneously with the placing - the section being continuously brought up to level in order to keep the lorries running - and in that case the degree of settlement experienced later may be negligible. It is desirable for some considerable exposure to rough weather to occur in order to consolidate the rubble, and this is particularly true of all the sections in which dumping and tipping is permitted.

2.7.6 It will be noted that the above allowance of 10% does not need to be increased to take account of any material displaced (item iii), because any such displacement will occur while the core is being built up to level, and will therefore be automatically made good. This extra volume of stone does however represent a straight loss to the contractor, which he must be required to assess at the time of tendering and allow for in his schedule rates. Whether or not it is reasonable to place all the risk for assessing the excess quantities on the contractor depends very much on the particular site conditions and upon the thoroughness with which the soil investigations have been carried out. If variations along the alignment of the breakwater are substantial, and if soil investigations are limited in extent (as may well be in the case for a particularly exposed site) the Engineer must decide whether the quantities at risk are an unreasonably high proportion of the total quantities, and in such cases a totally different basis of measurement must be considered - i.e. measurement on the basis of tonnage of material actually delivered to site and incorporated in the permanent works. It should be borne in mind that if the risk element is too high many of the tenderers, including the most responsible ones, may refuse to tender on any other basis than material actually used.

2.7.7. In the design and specification of wave walls and other structures to be placed on the breakwater it is desirable that note should be taken of the settlement problem. The contractor should be discouraged from carrying out the concreting too early in the programme, and we suggest incorporating a definite proviso that work shall not be started on this work at any point until the core and basic armouring for the whole structure is complete. In the case of very long breakwaters this work could be started in sections however. It is quite sensible for the wave wall and the completion of upper armouring to be carried out as the final tidying-up operation before the handing over of the breakwater, or of a substantial section of it.

2.8 Contractor's Risks

2.8.1 With regard to the supply of rock, a clause in the Specification normally allows the Contractor to obtain the rock or stone required for the Works from any source or sources approved by the Engineer, with the proviso:-

"Notwithstanding any approval given by the Engineer, it shall be the responsibility of the Contractor to satisfy himself that the source is capable of supplying rock of the sizes, quality and quantity required. The Contractor shall be responsible for obtaining all licences, wayleaves, permissions, etc., necessary for extracting rock".

2.8.2 Responsibility for the making good of any storm damage is covered in general terms by the normal contractual clauses (e.g. ICE Conditions of Contract clause 20).

"In case any damage ...... from any cause whatsoever ..... shall happen to the works ..... the Contractor shall, at his own cost, repair and make good the same ....."

2.8.3 It is considered that this clause gives the Employer all the protection that is reasonably needed in a contract in which payment is related to finished work measured in-situ. Where measurement is on the basis of material actually delivered, it is considered necessary to amplify the damage clause to the effect that if, following the occurrence of storm damage, the contractor fails to restore the displaced material to within the designed profile, a deduction will be made from the quantity of stone qualifying for payment in respect of any material lying outside the theoretical designed profile of the structure. It is acknowledged that in practice this may be a difficult clause to apply fully, especially in conditions where sounding is difficult, but some such provision is necessary to encourage the contractor to exercise proper care in controlling the work. In addition, it is desirable to limit the risk by specifying that the contractor must not allow the **unprotected** core to progress more than a stipulated distance ahead of protective coverage with secondary armour, and that the secondary armouring should similarly be covered by the outer layers within reasonable time.

2.9 Concrete Armour Units

2.9.1 Time and space do not allow us to cover in any depth the specification for artificial armour units, which, in any case, will have been dealt with more fully in other papers. All we need to say in this paper is that the concrete units should be made of concrete fully conforming to the specification for concrete in the marine environment, and that the same general rules apply to the handling and placing of artificial armouring units as apply to rock armour. As for the special instructions regarding individual placing and the integrity of the individual units during and after placing - these are particular to the specified units and must be individually covered in the specification. The sizes of the layers of stone under the concrete armouring must be determined in relation to the voids in the armouring, and the remainder of the structure will follow the normal rules for a rubble mound.

Discussion on Papers 11 and 12

A. SHAW *(Archibald Shaw & Partners)*
Delegates may be interested in the methods
adopted to build a small (200,000 t) rubble-mound
breakwater in the Clyde. The client required the
breakwater quickly, a matter of months rather
than years, so decisions were made using
'engineering judgement'. We also worked rather on
the lines suggested in Paper 10.

The main problem was to build the breakwater
on a rather soft sea bed at a price which would
make the project viable.

There was a large quarry that produced
excellent rock which could be delivered direct
into 1000 ton bottom-dump barges. The passage by
sea (about 70 miles) was relatively easy.

Rock could be dumped in situ, with only the
upper levels requiring to be double handed.

The amount of settlement that would take place
was difficult to predict, so it was decided to
spread a wide mattress of quarry run rock 3 m
thick on which the breakwater would be built.
We tried to place a fabric membrane (geotextile)
over the sea bed but were defeated by the tide,
which carried it away however carefully it was
anchored with rock.

Instead, we used mats of steel reinforcement
to provide an initial base and prevent local
penetration of the rock into the sea bed. These
mats were comparatively easy to lay and have
proved effective - the settlement of the structure
being less than expected.

Design of the armouring and underlayers has
been made to use the proportions of rock sizes
produced by the quarry. Rocks for the armouring
are stockpiled.

Further settlement of the breakwater is
anticipated and allowance will be made for this.

J. S. MOORE *(French Kier Construction Ltd)*
The view emerging from this conference is that
there is a great deal of uncertainty about the
stability and the durability of rubble-mound
breakwaters.

With regard to improvements in the design of
breakwaters, I would like to emphasize that there
should be no reduction, but hopefully an improvement in their constructability.

If a breakwater, such as that just completed
at Douglas Harbour, was designed to survive in a
hostile environment, then it is likely that it
will be constructed in that same hostile
environment (see Fig. 1).

The aim must be that the contractor can
batten down the hatches, ride out the storm and
then start from where he left off as soon as the

storm has abated. Limitations to forward progress
due to the need to repair damage should be
designed out as far as is practical.

Figure 2 shows the completed breakwater while
Fig. 3 shows an earlier stage with the concrete
capping and wavewall under construction,
retreating shorewards from the roundhead. Note
the 2½-6 ton rock underlayer placed against the
capping and the overlying 23 t stabits rising up
to a crest level of +13, 1½ m above the top of
the wavewall. These are shown to illustrate
the two-level approach to the construction.

Primary construction was carried out from the
+3.5 level, 9½ m below crest level and just
above H.W.O.S.T. Fig. 4 is an aerial view showing
the sequence of operations in that primary stage.

The breakwater core is of 0 - 2½ ton quarry
run granite rock, end-tipped into 20 m water
depth, spread and levelled by dozer. The core
material assumed a profile of 1 in 1 which over
a period would be worked down to the design
profile of 1 in 1½ by swell or storm.

Final profiling, carried out by dragline, was
not attempted until we were ready and able to
protect it. This immediate protection was
absolutely vital and once a number of grid lines
had been approved they were immediately covered
by a single skin of the 2½ - 6 ton underlayer
in the tidal and supra-tidal zones.

The next activity was to complete the double
thickness underlayer working from the soliplate
upwards.

The final activity in the primary stage was
placing the 23 t stabits, from soliplate upwards
and laid back at an angle of 45°. Although the
working level was +3.5 these stabits were
continued up to +7. or +8 forming a temporary
wavewall to provide shelter in storm conditions.

During the progress of the works the sections
most susceptible to damage were as follows:

(a) Tipped core: Redistributed and spread about
    but in these depths of water not much spread
    outside the design profile.

(b) Running surfaces: Frequently damaged but
    relatively easily (if not cheaply) made
    good.

(c) Profiled core: Damaged by storms. Compara-
    tively easy to make good, provided that
    there was no shortage of armour sized rock.

(d) Double skin underlayer: Less easily damaged
    than (c), but when it occurred the repair
    was slow, costly and had a marked effect on
    progress. Jumbled core and underlayer

Breakwaters—design and construction. Thomas Telford Ltd, London, 1984

159

Fig. 1. Hostile conditions during construction of breakwater

Fig. 2. Completed breakwater at Douglas Harbour

material washed down onto the laid back stabits. Larger rock, if available, could have reduced the incidence of this type of damage.

Considerations of factors such as these by researchers and designers may result in more constructable and therefore more durable breakwaters. I believe that Douglas Harbour was a step in the right direction.

T. WILSON *(Charles Brand & Son (N.I.) Ltd)*
Reference during the conference to quarry suitability has mainly been conferred to the contractor operated quarries, with little comment made on the fairly common usage of local commercial quarries.

For the construction of Bangor breakwater in Northern Ireland, armour stone in the 6-10 t range was supplied by local quarries working fresh olivine basalt deposits with a S.G. which varied from 2.70 to 2.85. Despite many variations to the blasting techniques the yield never exceeded 7% and had an average of around 5%. Bearing in mind the experience at Port Talbot, where more than 30 quarries were used, and the close similarity of rock in the north of

Fig. 3. Early stage of construction, with concrete capping and wavewall under construction

Fig. 4. Aerial view showing the sequence of operations in the primary stage of construction

Ireland to the rest of the UK, this yield figure may be fairly representative of the UK as a whole.

A yield of this size will, in many cases, ensure an over production of quarry run material. In the past this has not been a problem for the breakwater project, as commercial quarnes simply crushed the material to produce aggregates for their normal business. With the Bangor project, however, the current economic recession has not provided a market for this excess material with the result that the quarries either reduced the rate of production of armour or ceased production completely. Bangor breakwater is currently being

constructed using SHED units above mid-tide level in place of the unavailable rock armour.

In view of the experience at Bangor it is, perhaps, necessary to extend pore-tender investigations of the quarries to include a forecast of the local economic climate to establish whether it can sustain such a boost to the output of normal quarry aggregates.

W. BEAZLEY (Rendel, Palmer & Tritton, Paper 11) Could M. Maquet elaborate on the placing of large 50t tetrapods (para. 72 of Paper 11) to get optimum stability with density in the second layer? As it has been our experience in placing

7,700 large tetrapods (in a total of 33,500 tetrapods) that the word random in 'Sofremer' and 'Rendel, Palmer & Tritton' specifications cannot apply, after 'hook positioning'. In other words large tetrapods must also be orientated accurately.

J. BERRY *(Bertlin & Partners)*
Can poor quality rock be used in the core of a rubble-mound breakwater? The core in its final condition, is protected by the outer layers from most causes of deterioration and on projects where a large volume of poor quality rock is available it may be uneconomic not to use it. I understand that chalk, which comes under this description, was used for the core at Antifer and I wonder if now, following completion, any information on its performance is available.

Sluicing of rockfill is carried out for dam construction and sometimes for breakwater construction in order to wash out fines. I have seen it used, but I am not sure whether it is effective or worth including in a specification. Could the Authors comment?

I agree with the Authors that armour stone over 2 t weight should be placed and not dumped. I know of a case where a breakwater toe was constructed by dropping a 16-20 t rock from above water level to a depth of some 20 m below chart datum and I cannot believe that it finished up in its planned position. A further point about armour is that of the correct placing of elongated rock, say 1:2½ in dimension. It is best placed as a header, rather than on the flat. Is it possible to enforce this requirement by suitable provisions in the specification?

I have found it difficult to draft suitable specification clauses to describe permissible tolerances for a multilayer rubble-mound break-water. The only solution is to incorporate a drawing of the cross-sections showing them.

I feel that the specification should include the requirement that approved samples of all weights of rock armour should be mounted on plinths at a suitable position on the site so that disagreements between the Engineer and the Contractor as to what is acceptable and what is not acceptable are minimized.

P. HUNTER *(Sir Alexander Gibb & Partners)*
Most of my comments arise from comparing Paper 11 with specification requirements we have developed over a variety of breakwater projects in recent years, with an emphasis on achieving economies without impairing the structure. I refer to the paragraph numbers in the Paper:

2.3.3. The drop test is eminently practical and can easily be used on site, and the use of a constant height of drop imposes correctly a higher strength requirement on the larger rocks. Is the required number of tests specified? We would normally specify a minimum number which would be increased if necessary. Presumably, a tiny number of rocks are dropped in this way?

2.4.4. <u>Grading of core</u>. In a previous discussion Mr Hedges asked whether a limit should be placed on the minimum size of material in the core. The same question can apply to the upper limit.

I would suggest that a smoothly graded material with a D_{50} corresponding to a weight

between 12 and 250 kg is adequate in the finished structure, and the contractor should be allowed to select the maximum size to suit the season and his quarrying and the construction methods. With regard to the lower limit, if nothing less than 10 kg is permitted, this might be rejecting 20% of the quarry output, and inevitably there will be a significant percentage below this weight anyway. A more realistic and economical requirement is to limit rocks of less than 1 kg to less than 5% of the total weight.

2.4.9. <u>Single-sized main armour</u>. ±25% on the weight is only ±8% on diameter. This can be relaxed without any perceptible effect on armour stability, at least to ±35% or 40% by weight.

2.4.10 <u>Underlayers</u>. Although a large-sized underlayer is more stable during construction, this benefit is largely lost if it leads to a need for an additional smaller layer beneath it. Any surplus larger stones, in the unlikely event of them occurring, can probably be placed in the core. Reduction of the total number of layers, leading to a simple structure without tight underwater tolerances, is a high priority.

2.5. <u>Placing stone</u>. Although normally all slopes within the breakwater are parallel to the armour slope, typically 1:1.5, the core can be placed at a natural slope if the contractor wishes, with a thicker sublayer to achieve the specified slope at the next interface. This reduces the need for accuracy in construction of the core, at what can be the small expense of substituting coarser material for core material in a small sector of the section.

2.6.7. <u>Additional checking of quantities of material by weight</u>. I am not sure how this can form part of a contract specification without superseding the measurement of the as-built structure.

2.7.4. If all dimensions on the cross-section are increased by 10%, there is an increase of over 20% in rock volumes, with a consequent enormous additional cost.

If only the vertical dimensions are increased, the 10% increase in slopes implies a 25% reduction of acceptable wave height (I am care-fully avoiding the term design wave). I would suggest that contractors have a sufficient challenge if required to build exactly what is on the drawings.

Settlement of the order of 10% does not, in our experience, occur except perhaps with materials of very much poorer quality than those which the requirements of this Paper will keep to. Such a settlement might well be described as a failure.

2.8.3 Why is it necessary to put one parti-cular limit on the contractor with regard to his construction method? Is this not best left to the contractor?

These suggestions are not intended to improve the quality or stability of structures conforming to what is undoubtedly a thorough and tested specification, but provide economies, where these are possible, without impairing the structure's performance.

P. G. FOOKES *(Consultant; Queen Mary College)*
I would like to comment on sections 2.3, 2.4 and 2.5 with which I find myself in agreement.

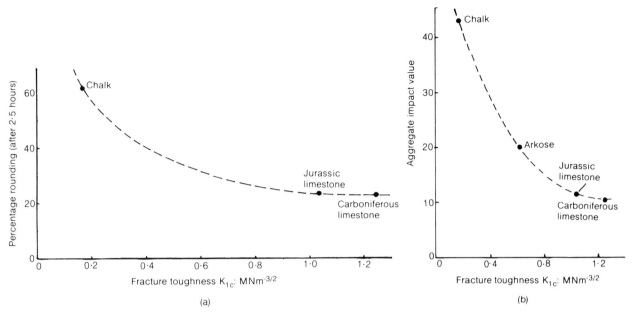

Fig. 5. (a) Percentage rounding against fracture toughness; (b) aggregate impact value against fracture toughness

I want to make a plea for giving limits to various test requirements, to be met by rock being selected for use in the breakwater. This gives teeth to the Engineer if he has borderline material and enables the Contractor to know fairly precisely what is required of him.

Paper 3 (Table 6) gives some suggested limits but only as a guideline; each structure deserves to have its own test and other requirements to be specifically designed for it. Aggregate Abrasion Value and the Los Angeles test are not specifically suggested in the guidelines. The interpretation of these tests is sometimes problematical in the marine context and is currently being investigated. It is hoped to be able to make some positive recommendations on them in the not too distant future.

Speaking generally, engineering tests are sometimes of limited value in predicting the behaviour of rock. Armour blocks, for example, depend on their full size to fulfil their role. Macro-flaws such as cracks and mineral veins can reduce the effective strength of such units considerably. Engineering tests may indicate the ultimate strength but are generally not capable of identifying the reduction in strength arising from flaws. In practice the majority of solids show only 1/100 to 1/1000 of their theoretical strength.

The degradation of rock has been discussed in Paper 3, the differences between the three deterioration processes (abrasion, spalling, catastrophic breakage) being essentially one of scale; sand grain impact sets up the same kind of pressure as rock impact but with a smaller load, and degradation is therefore surface rather than the whole block. Salt attack, for example, can be considered as setting up descriptive pressures from within cracks and pores rather than applying an external pressure.

It is suggested by the research at Queen Mary College, that fracture toughness of the rock, as measured by the stress intensity value K_{IC} (ASTM E399-78a) is perhaps the best indication of the combined cohesive strength of rock minerals and grains against which disruptive forces operate. The advantage of using such a parameter is that it reflects the overall quality of the rock rather than any individual strength affecting factor. Thus a calcareous cemented sandstone and a quartz cemented sandstone of equal porosity, may have similar values of Young's modulus. Fracture toughness takes into account all variables and would indicate the stronger rock on the basis of porosity, strength, cement-grain bond and any other significant parameters present in the rock.

Preliminary estimates of K_{IC} values have been obtained for various rock types. Fig. 5 gives examples of the work in progress and relates fracture toughness by the method with the single notched beam, with percentage rounding in the laboratory machine and also with Aggregate Impact Value, a standard mechanical test for aggregate. In summary the research looks to have some promise but is not ready for commercial application yet.

W. A. PRICE (Hydraulics Research Station Ltd)
With reference to the graph of percentage roundness against damage coefficient, could you explain further what damage coefficient means and how it was measured.

I measured the force required to pull out dolos, tetrapods, stabits, etc. from a slope. With dolos, I progressively filed the edges off the edges to imitate wear. I found no difference in the pull out forces but the voids were reduced. I do not know but I would suspect that reducing the voids would decrease their hydraulic stability. Hence such structures would get weaker with age.

J. R. TAYLOR (Wilton & Bell Property Ltd, North Sydney, Australia)
I was most interested in the earlier description of the work at Ras Lanouf Harbour, in that my firm has recently been responsible for two major

breakwaters in the Gulf of Oman.

The two breakwaters totalled some 4 km in length into 15 m of water for which the contractor worked 24 hours a day, 13 days a fortnight, and which contained upwards of three million cubic metres of stone filling armoured with almost 40,000 dolosse units.

Stone sizes were specified to be: core material 1-500 kg, underlayers of 500 kg to 1 t and 1.5-4 t, with 4-6 t stone at the toe of the dolosse slope.

I wish to make two specific points: First, a plea to consulting engineers and site staff for flexibility in the specification and its interpretation.

In our case, the quarry production from a quarry that did not have the benefit of earlier investigation as to yield, was deficient in the 500 kg to 1 t size and breakwater construction was falling behind programme because of the lack of this grading. With deteriorating progress it was decided to take the 500 kg size out of the core and place it in the underlayer and to increase the upper limit size for the underlayer. From the original specification of 1-500 kg and 500 kg to 1 t, the core and underlayer were changed to 1-300 kg and 300 kg to 1.5 t respectively. This more than doubled the quarry output and underlayer material with corresponding increase in breakwater construction progress.

In addition, specific tolerances were allowed in the slope of core material. The change from a tipped natural slope of core to the required outer armour slope of 1 to 2 was achieved by thickening the first underlayer progressively with depth. Natural tipped core slopes under calm sea conditions were generally at 1 to 1.3. However, the contractor was given the option of flatter core slopes up to 1 to 2, where sea conditions produced such slopes or plant choice allowed, so long as the minimum underlayer thickness was maintained with consequent savings to the client and increased rate of construction with the reduced underlayer size requirement.

Secondly, I commend the Ras Lanouf concept of weighed samples of each specified stone size being placed on display on plinths both at the quarry stock pile area and at the resident engineer's office as a constant reminder. We arranged just this with each size labelled. Constant daily observation of each size and optimum shape is a great help in continually reminding both resident engineer and contractor's staff, foremen and inspectors of each grading size

Experienced quarry men recognize each required stone size easily but daily reminder is necessary for all supervisory staff if standards are not to drop over the many months of construction.

A. H. BECKETT (*Sir Bruce White Wolfe Barry & Partners*)
There appears to be an anomaly in our attitude to armouring of rubble-mound breakwaters. If we recognize that the sea in its furious moments wears down the hardest rocks, and flattens the sea slopes of breakwaters, why do we choose to armour on flattish slopes of say 1 to 3 with the hardest and most durable rock, yet consider light and fragile concrete units as capable of

armouring slopes as steep as 1 in 1½ of areas of severe sea attack?

J. F. MAQUET (*Paper 11*)
In reply to Mr Beazley, the word random is sometimes misused or misunderstood when speaking of placement of units. Only rocks of small size and quarry run are actually randomly placed.

For each type of unit a placement scheme must be followed in order to achieve the specified number of units per unit area. When dealing with cubic units, the word random refers more to the appearance of the armour which must not look like brickwork rather than to the placement itself, although the orientation of units could be random in this case. With multilegged units we agree with Mr Beazley on the necessity of accuracy not only in placing but also in orientation of the units since it has an effect on stability.

P. G. R. BARLOW and M. G. BRIGGS (*Paper 12*)
With regard to Mr Berry's questions about control, we agree that it is difficult to achieve the desired result when we specify the way in which rock should be placed and the acceptable tolerances. In the end it all comes down to the capabilities of the man on site - the Engineer's Representative - guiding and controlling the contractor. It seems to be possible in practice to achieve satisfactory results and judging from one of Bertlin & Partners' breakwaters, which we have recently been able to study closely, their Resident Engineers have been no less successful than ours in achieving good quality work.

Mr Hunter has raised some interesting points: Regarding the frequency of drop tests (para. 2.3.3), this is very much dependent on results obtained. If 100% success is achieved in each test, the frequency can be very low - just an occasional check sample. However, if a failure occurs, tests must be more frequent and must be continued at the higher rate until they are seen to be a virtual formality.

In practice we would not dispute the acceptability of 5% of stone less than 1 kg in the core, provided of course that it does not all appear in concentrated areas. The real point is that unless from the beginning of the contract the contractor sets up proper arrangements to screen out the very small material, the result will be haphazard, with constant arguments on site, and actual deliveries containing 5-25% of smalls. We would regard 5% as the limit of discretion on the part of the site supervision.

Regarding the allowance for settlement (para. 2.7.4.) we admit that we are not at all sure that the suggested clause provides the right answer and we are very much open to suggestions for improvement. The 10% figure is of the right order, but in ordinary open sea conditions, and where stone is end-tipped so that the passage of trucks continually agitates the material already placed, this settlement usually takes place during construction and is probably never noticed. A similar structure built with floating plant during a calm season will undoubtedly show substantial settlement during the first stormy season. The purpose of this clause is to warn the contractor to make some allowance for this factor so as to achieve the objective of having the structure fully up to profile at the end of the maintenance period. Unless it is fully up to profile it is very difficult to make it so, for the degree of settlement is probably much less than the thickness of one unit of primary armouring; so you either have to remove the armouring and make good the layer underneath or build up very much above profile.

It is of course possible to put the onus entirely on the contractor with some such clause as: 'The intention of the design is that the crest levels and general slope profiles should finish up no lower than the lines and levels indicated on the drawings at the end of the Maintenance Period. The contractor shall be entirely responsible for taking the measures necessary to achieve this objective, either by making due allowance for such settlement during construction or by restoring the structure to level before the issue of the Maintenance Certificate'.

With regard to para 2.8.3. why is it necessary to specify that the contractor must not leave an underlayer unprotected more than a certain distance before it is covered by the outer protection? In our experience it is surprising how often the contractor's man on site needs some outside control to discourage him from taking unreasonable risks, and although any storm damage that may occur is usually at the contractor's risk, it is inevitable that progress will be delayed

in reinstating the damage, so that the employer has a direct interest in the matter. More importantly, there is the question of insurance, and the existence of this clause makes it easier for the contractor to obtain insurance cover at reasonable rates in the expectation that the Engineer's Representative will be monitoring the contractor's work and thereby keep the insurance risks within defined limits.

We were gratified that Professor Fookes was in general agreement with our Paper. We do not in any way disagree with his view that the specification is incomplete without figures for the test requirements. However, we think it inappropriate to give hard and fast figures in a general paper; these should be given when the site and the material that is obtainable within a practical distance of the site are known. At that time we could do worse than to come back to Professor Fookes and ask him what figures to use!

A. B. POOLE, P. G. FOOKES, T. E. DIBB and D. W. HUGHES *(Queen Mary College)*
In reply to Mr Price, the damage coefficient used by us is the proportion of unstable or substandard blocks (rounded, degraded or fractured), expressed as a percentage of the total number of blocks in the given study area, usually 100 m^2. Further discussion is given in the introduction to Paper 3 and in ref. 3 of the Paper.

13 Monitoring of rubble mound structures

O. T. MAGOON, V. CALVARESE and D. CLARKE, US Army Corps of Engineers

---

Rubble mound structures are one of the enigmas of coastal planning and coastal engineering. In spite of major theoretical and hydraulic laboratory research and study, the past decade has seen many costly problems associated with rubble mound structures. The designers of these structures were generally experienced and capable, implying that the design of a rubble mound structure, particularly in deeper water and exposed to higher seas needs additional study effort. In this conceptual framework, monitoring of rubble mound structures remains extremely important; both as a means to insure that the structure is performing an intended purpose and also to provide basic information and observations on prototype structures to serve as a guide to designs at other locations and future research.

The monitoring of a rubble mound structure must consider the original design and construction as well as the forces and factors which produce changes in the structure. Almost all coastal structures receive some monitoring, and a vast amount of information has been published on case histories which include some references to monitoring. Langdon and Fowler (Ref. 1) present good examples of monitoring using conventional surveying techniques. Magoon, Sloan and Foot (Ref. 2) present an overview of major structural changes experienced and monitored at Nothern California coastal structures. Repeated surveys of large concrete armour units from the structure cap are effective in determining major concrete armour breakage or serious structure deterioration in areas of the structure above water. For example, inspections of the Crescent City Harbor breakwater, California in 1982 indicate that more then 25 percent of the 42 ton unreinforced concrete dolos units are broken; many into more than two pieces.

As major coastal structures extend well below the still water level, it is desirable to monitor and evaluate the portions below water as well as the portions of the structure above water. However, due to the expense of underwater inspection, often combined with rough seas and poor underwater visibility, many inspections (which consist of the only monitoring a structure will receive) are made only from its crest or cap. These inspections cannot ordinarily detect breakwater problems in early stages, but generally only problems in advanced stages. This is because the observer's field of view is limited to a relatively small viewing angle. Additionally, unless the structure is in an area of large tidal range, portions of the structure normally underwater cannot be observed.

A better approach for a general view is an inspection and monitoring both from the structure and from a small boat at close range. Kluger (Ref. 3) gives a pratical discussion of simple techniques for monitoring from the water surface by comparative photography.

Aerial photography combined with photogrammetric techniques using plotted scales of say one-inch equals five feet, combined with adequate ground control represent an excellent tool for monitoring changes.

Underwater monitoring includes soundings taken by hydrographic surveying means as well as taken with land based equipment such as portable cranes, hydrographic surveys and helicopter-based surveys. Side Scan Sonar is a promising technique and will be discussed later.

Underwater inspection by trained observers is an excellent method of monitoring rubble mound structures. This method is very time consuming and costly as wave conditions and poor underwater visibility may hamper the determination of underwater positions. Detailed underwater surveys have been made, however using divers and controlled horizontal and vertical positions for the rehabilitation of the Diablo Canyon Breakwater north of Port San Luis, California. In this case, locations of displaced and broken Tribar concrete armour units were carefully plotted.

Inasmuch as the condition of a rubble mound structure is of critical importance to the responsible authority, the structure designer should develop a monitoring program concurrent with design, so that those responsible for operation and maintenance may know in advance

Breakwaters—design and construction. Thomas Telford Ltd, London, 1984

167

QUADRIPOD VERTICAL DISPLACEMENT

(feet)	Number of Quadripods				
	June 63 to June 64	June 64 to June 65	June 65 to May 66	May 66 to April 67	April 67 to Dec 68
0 - 0.05	21	21	21	22	22
0.06 - 0.10	8	5	12	5	9
0.11 - 0.20	6	6	5	2	5
0.21 - 0.40	5	2	3	2	
0.41 - 0.60	1		1	1	
0.61 - 0.80				1	
0.81 - 1.00				1	1
Greater than - 1.00	2		1	1	1
	43	34	43	35	38

what corrective actions need to be taken, and additionally, what information should be developed to increase our knowledge of rubble mound structures. As mobilization of the large construction equipment necessary to effect repairs is often very costly, recommendations as to maintenance and monitoring to insure the integrity of a structure should be as specific as possible. For example, the number of broken or displaced armour units that may be tolerated at a specific location and on the total structure could be specified. Maintenance of appropriate stockpiles for small repairs, based on monitoring may be cost effective.

The most generally accepted factors which are associated with damage or erosion of a rubble mound structure are waves, tides and currents. A great number of factors which affect the still water level are of course important. In attempting to monitor the waves at a structure, it is important to remember that most deisgn waves for coastal structures are calculated for the condition before the structure was built. Thus any attempt to compare prototype wave heights and water levels with design conditions should carfully consider suggestions from those involved in design and hydraulic modeling. Measurement of water levels are generally made at some distance from the desired area of the structure and thus may show significant differences from actual values at the structure. Unfortunately, recording devices may fail to operate during severe conditions so that desired extreme values may not be obtained. Many locations may have specific problems caused, for example, by earthquakes or tsunamis and these must be provided for on a case by case basis.

Three examples of American breakwater monitoring will be described. More detailed information on these examples is available both in published and unpublished form. Monitoring programs to be presented are Santa Cruz Harbor West Jetty, California, Weymouth and Magoon (Ref. 4); Manasquan Inlet Jetties, New Jersey; and the Cleveland Harbor Breakwater Rehabilitation, Ohio, Pope, J. and Clark, D. (Ref. 5)

Santa Cruz Harbor, California.

Santa Cruz Harbor is located on the northerly end of Monterey Bay about 65 miles south of San Francisco. The harbor was formed by the development of Woods Lagoon, a shallow pond near the eastern boundary of the City of Santa Cruz. The harbor includes an entrance channel to Montery Bay and the Pacific Ocean, protected by two rubble mound jetties. The west jetty is about 1200 feet in length and the outer portion is protected by 28-ton concrete quadripods. For the quadripod sections, the selected design called for two layers of quadripods (pell mell) placed on a 1 on 2 slope along the trunk and on a 1 on 3 slope around the conical head section. The quadripods were backed by a concrete cap, essentially 18 feet wide and 10 feet thick to prevent displacement of armor units from the cap by overtopping waves.

Upon completion of the structure, detailed surveys were conducted to serve as a basis for measurement of future displacements. these surveys consisted of establishment (horizontally and vertically) of 5 standard brass disks in the concrete cap. Steel pins were set in each corner of concrete cap pours and in 44 numbered quadripods throughout the upper layer. Repeat measurements of all monuments and pins were taken periodically and of selected quadripod pins after stormy periods. Horizontal and vertical distances are measured from the cap by standard surveying techniques. Vertical displacements recorded during a 4-year period are summarized in the table following. All quadripods above water are also identified in aerial photographs of the seaward portion of the jetty.

With the exception of a few quadripods near the jetty head, no major movements have occurred. The maximum cap settlement is about 0.1 foot, near the seaward end. Quadripods continued to move for five years', however, shoaling seaward of the structure has reduced wave heights. The armour units appear to be consolidating with settlement, the individual units tend to interlock thus providing for maximum stability. In general, movement of a few units without breakage is not indicative of failure in the structure. During the two decades since construction, some loss of the explosively driven steel pins have been observed. Aerial photography and inspection from the cap confirms that the structure is stable.

Manasquan Inlet Jetties, New Jersey.

The Manasquan Inlet Navigation Project is located on the Atlantic coast of New Jersey, approximately 26 miles south of Sandy Hook and 23 miles north of Barnegat Inlet. The inlet provides the northernmost connection between the New Jersey Intracoastal Waterway and the ocean. The first attempt to stabilize the inlet was begun in 1881.

The existing Federal project at Manasquan River includes parallel rubblemound jetties approximately 1200 feet long, spaced 400 feet apart, protecting a channel 250 feet wide by 14 feet (mean low water) deep. Following project completion in 1933, up to about 1970, the jetties suffered recurring storm damages which necessitated frequent maintenance and repair. At present, (December 1982), both jetties have been rebuilt using 16-ton dolos reinforced concrete armour units. The jetties, entrance channel and adjacent beaches are being monitored. This represents the first application of dolosse on the U.S. East coast.

The monitoring program developed has three basic objectives:

1. Jetty Stability. Evaluate the performance of the dolosse armor units in maintaining the structural stability of the jetties. This includes the structural integrity of the dolosse themselves.

2. Effects on adjacent beaches. Evaluate the potential effects of the rehabilitated jetties on longshore sediment transport in the vicinity of the inlet.

3. Inlet Channel stability. Evaluate the effectiveness of the rehabilitated jetties in maintaining a stable inlet channel cross-section.

Only the first objective will be considered in this paper. The tasks required to carry out the jetty monitoring program are listed below.

1. LEO. The littoral environment observer (Berg, Ref 6) will use the seaward end of the completed south jetty as an observation site, with twice daily visits. Measurements will include: wave direction, wave height, wave period, wind speed and direction.

Present plans do not call for routine measurement of littoral current speed and direction, nor for measurement of adjacent beach slopes. However, the LEO will make daily visual inspections of the dolosse in order to discover any obvious breakage or

movements which otherwise might not be detected until the next scheduled jetty survey.

2. Wind Speed and Direction. Measurement of wind speed and direction is obtained every three hours at a nearby location.

3. Wave Gage. A Datawell "Waverider" wave gage is installed at a location about 0.8 miles northeast of the seaward end of the north jetty. The depth at this location is about 50 feet. Wave data will be analysed to determine prevaling wave climate and critical values.

4. Tide Gage. A recording tide gage is installed in the inlet.

5. Tidal prism measurement will be made once annually.

6. Side Scan Sonar. A side scan sonar survey will be performed in Manasquan Inlet once during the first year of the program, using the Klein Model #531 unit with an operating frequency of 500 kHz. The primary objective of this task is to evaluate whether side scan sonar techniques are capable of resolving the location and orientation of the submerged 16 ton dolosse on the inlet channel side of the jetties. Because of lack of experience in this specific application of side scan sonar, the work is considered experimental in nature. Consequently, the details of the survey regarding number of survey lines, distance from the jetties, navigation control, and toe depth, may be modified during the course of the survey. If the results of the initial survey are satisfactory, one additional survey will be performed annually during the monitoring program. An emergency survey would be performed following a major storm in order to assess displacement of submerged dolosse.

10. Aerial Photography. Aerial photographic flights are scheduled quarterly to coincide with the surveys. Coverage will extend about six miles north and south of Manasquan Inlet. Black and white imagery will be obtained from an altitude of 2400 feet, resulting in a contact scale of 1" = 400'. An additional low-altitude flight line will be made along the axis of the inlet channel. The imagery from this flight line will be used in the photogrammetric analysis of dolosse stability described below. (Task 12)

11. Dolosse Stability Surveys. The purpose of the dolosse stability surveys is to accurately establish the location and orientation of selected dolosse on the two jetties and through subsequent surveys to measure any displacements which occur. The timing of these surveys will be quarterly and scheduled to coincide with the beach and hydrographic surveys. The procedure will involve establishing nine survey lines on the south jetty, and eleven on the north jetty. The lines will be spaced 100 feet apart along the jetty trunks where dolosse have been

TASK	MONITORING OBJECTIVE ADDRESSED		
	Jetty Structural Stability	Shoreline Response	Inlet Stability
1. LEO (Littoral Environmental Study)	X	X	X
2. Wind Speed & Direction	X	X	
3. Wave Gage	X	X	
4. Tide Gage	X	X	X
5. Tidal Prism	X		X
6. Side Scan Survey	X		X
7. Inlet Hydro Survey			X
8. Beach Surveys (on/off shore)		X	X
9. Sand Samples		X	X
10. Aerial Photography	X	X	X
11. Dolosse Stability Surveys	X		X
12. Dolosse Stability - Photogrammetry	X		X
13. Site Inspection	X	X	X
14. Project Management	X	X	X
15. Data Reduction/Reports	X	X	X

placed and will extend radially from the jetty heads at 45° intervals. Along each survey line, two dolosse will be selected for monitoring, one near the structure crest and one near the waterline. Thus 18 dolosse on the south jetty and 22 dolosse on the north jetty will be monitored. At least one permanent survey point will be established on each of the selected dolosse. By using existing horizontal and vertical control established for project construction, the x, y, and z coordinates of each selected survey point will be determined. Coordinates of each survey point will be tabulated after each survey, and compared to the preceding set of coordinates. Where differences are evident between surveys, the magnitude and direction of displacement of each dolos will be calculated.

12. Dolosse Stability - Photogrammetry. In addition to the standard survey techniques just described for monitoring potential dolosse movement, an experimental program using photogrammetric techniques has been implemented. The purposes of this task will be to determine (a) if photogrammetric techniques are capable of detecting displacements as small as are detectable by standard survey methods, and (b) if so, can an equal or greater number of dolosse be monitored for the same cost. As part of this task, a number of first order control points will be established in close proximity to the

completed south jetty. These points will be identified on the low-altitude flight line photography (as described previously in Task 10) and used as control for producing a template of projected end area of all visible dolosse. The template from the initial flight then defines the base condition against which possible future dolosse movements would be evaluated. If photogrammetry is found to compare favorably with standard survey techniques for the dolosse monitoring then an additional experimental task would be incorporated into the program to evaluate photogrammetry as a technique for monitoring sediment volume changes on the beaches adjacent to Manasquan Inlet. Actual surveys at a scale of 1" = 5' will be displayed at the conference.

13. Site Inspection. Site inspection will be made quarterly by the project manager and a photographer. Activities to be accomplished on each visit include overall assesment of the condition of the jetties (south jetty is completed and the rehabilitation is in progress on the north jetty), visual inspection of the dolosse for obvious movement or breakage, ground level photography from selected locations on both jetties and the adjacent beaches of Point Pleasant and Manasquan, inspection of wave and tide instrumentaton installed at the site, and coordination with other personnel involved in the monitoring program.

14 & 15. Project Management/Data Reduction/Reports. These tasks are grouped together as they consist of overlapping activities, including project manager and technical support as required. Activities include preparation of scopes of work for contracts, contract supervision, data collection and analysis, coordination on program progress, and finally reporting of results.

The first set of detailed observations has been completed and the second set of detailed observations is set for early 1983. Comparison of armour unit movement and other results will be presented.

Cleveland Harbor, Ohio.

Cleveland Harbor is the major U.S. Lake Erie port on the Great Lakes. The East Breakwater is about 21,000 feet long. The easterly portion was constructed of laid up stone and completed by 1915. The breakwater provides protection for commercial shipping, a harbor of refuge and a berthing area for pleasure craft. The breakwater has a design crest elevation of +10.3 feet, a crest width of ten feet and side slopes of 1 vertical to 1.5 horizontal. The Eastern portion of the East Breakwater was rehabilitated in 1980 with two ton unreinforced concrete dolos armour units placed on a one vertical to two horizontal slope. A serial number and date were cast into each unit. The monitoring program relied heavily upon the use of aerial photography supported by ground control surveys to monitor above-water dolos movement, periodic inventories of individual armour units for breakage, and side scan sonar to document the underwater condition. Two submerged pressure wave gages were used to record the wave climate of the project site during the period of monitoring.

Many of the elements in the Cleveland Monitoring effort are similar to the those at Manasquan. Monitoring at Cleveland has been in progress for almost two years and some interesting information on side scan sonar and underwater inspections as well as practical results follow.

Side scan sonar surveys have been conducted twice since the initiation of the monitoring program, in April 1981 and in July 1982. For both surveys, a Klein Associates 500KHz unit was used to document the various coastal structures found in Cleveland Harbor including timber cribs, laid-up stone, rubblemound, and dolos breakwaters and numerous varieties of vertical-walled bulkheads. Lessons from the first survey caused improvements to the second survey: (1) a larger boat used (from a 17-foot to a 21-foot); (2) a short range electronic positioning system was used; and (3) operational procedure was documented for duplication. The dolos-rehabilitated East Breakwater was surveyed at least twice during each visit to determine the repeatability of the imagery. Although the individual dolosse are too small to be identified within the structure slope, the structure toe was distinctive, changes in the structure slope were discernible, and general repeatability between side scan sonar runs is possible. Apparently the wavelengths and signial processing of this device were not able to resolve the two-ton Dolos units.

Underwater (scuba) diving inspections were made in July 1982 along the trunk and the head. An additional inspection was made in September 1982 at the head. Visibility is generally poor in the vicinity of Cleveland Harbor, but the scuba diving surveys did provide some first-hand information on the underwater condition of the dolos cover. The results of this monitoring are presented in detail in the previously referenced paper on Cleveland Breakwater monitoring, however, dolos units are moving with the units near the still water line, experiencing the greatest movement. The movement has not diminished with time, but is associated with storm wave action. The average individual movement was about one foot over a period of one year.

Conclusions.

Monitoring of rubble mound structures yield valuable information both regarding structural integrity and also increases our knowledge of rubble mound structures. At locations discussed in this paper, concrete armour units have moved when subjected to wave action. In some instances breakage of armour units resulted. Movement of armour units continued over a number of years, rather than locking in place after the first winter.

Monitoring planning and associated operations and maintenance guidelines are sometimes neglected in the design of rubble mound structures. Implementation of well planned monitoring programs are essential to insure that a rubble mound structure serves its intended purpose, and also to provide prototype information to serve as a basis for improving new designs.

Acknowledgement is gratefully made to the Corps of Engineers, U. S. Army, for access and permission to use this study material. The view of the authors do not purport to reflect the position of the Corps of Engineers or the Department of Defense. The assistance of personnel of the San Francisco District Corps of Engineers is deeply appreciated.

References

1. Langdon, K. J. and Flower, B. L., Inspection and Maintenance of Breakwaters, Proceedings, Maintenance of Maritime Structures, the Institution of Civil Engineers, London, 1978. pp. 105-120.

2. Magoon, O. T., Sloan, R. L., and Foote, G. L., "Damages to Coastal Structures" Fourteenth Coastal Engineering Conference, Copenhagen, Denmark 1974, Coastal Engineering Research

Council, American Society of Civil Engineers, pp. 1655-1666.

3. Kluger, J. W., The Monitoring of Rubblemound Breakwater Stability Using a Photographic Survey Method, Abstracts 18th Conference on Coastal Engineering, Coastal Engineering Research Council, Capetown, South Africa, 1982, pp. 348-349.

4. Weymouth, O. F. and Magoon, O. T. "Stability of Quadripod Cover Layers" Proc. Eleventh Conference on Coastal Engineering, London, England, 1968 Coastal Engineering

Research Council, American Society of Civil Engineers, pp. 787-796.

5. Pope, J. and Clark, D. "Monitoring of a Dolos Armor Cover; Cleveland, OH",; Proceedings, Coastal Structures Conference, Washington, DC, 1983. American Society of Civil Engineers, In Press.

6. Berg, D. W., "Systematic Collection of Beach Data", Proceedings, Eleventh Conference on Coastal Engineering, London, England, 1968, Coastal Engineering Research Council, American Society of Civil Engineers, pp. 273-297.

Discussion on Paper 13

F. B. J. BARENDS (Delft Soil Mechanics Laboratory)
Research has to be directed to assess deformability of rubble mounds, not only to strength. Ultimate strength is usually mobilized after relatively large deformations. However, malfunction of a breakwater might occur before ultimate strength is mobilized. Due to wave impacts the entire structure responds generating cyclic deformation, which can be in the order of centimetres and which will cause erosion at the intergranular contact points and decrease of frictional resistance; consequently this affects the stability.

R. W. BISHOP (Cooper, Macdonald & Partners)
The photograph on the front page of the conference programme is of the Alderney breakwater. This is one of the oldest and most well known rubble-mound breakwaters, which in its most exposed position receives the full force of Atlantic gales; yet it has not been mentioned in any of the Papers.

In 1949 I presented a paper on repairs to the Alderney breakwater. The cause of damage to the breakwater had been the subject of much discussion. The initial requirement for and siting of this breakwater was dictated by military considerations. In the exceptional depth of water, which exceeded 135 ft at the seaward end, a mound-type structure would have required an enormous amount of stone and so a composite design was the obvious solution. Continual trouble has been caused by large waves breaking against the face of the wall with large volumes of water overtopping the wall and cascading on to the roadway behind, thus causing high impact forces and suddenly compressing air trapped in open joints. This could have been avoided by keeping the mound well below low water level or by bringing it high enough to cause the waves to break before they reached the wall.

Repairs in 1949 and since have consisted of filling the cavities with mass concrete, precast concrete, or various forms of bagwork, and repointing the blockwork in cement and sand with various additives. With differential settlement and the resulting cracking of joints as an inherent difficulty in such structures - remembering that the Alderney breakwater is typical of many in different parts of the world - one cannot help wondering if the use of an epoxy resin for repairs has been considered because materials have been developed which will bond to wet surfaces, have high strengths and moreover flexibility up to about 80%.

T. SØRENSEN (Danish Hydraulic Institute)
Breakage caused by compaction. It is impossible to place armour units in the densest possible configuration. As a result of this the armour layer will be compacted by the dynamic impact of the waves. Vibration is the best way to achieve maximum compaction of, for example, concrete and granular material. So the armour layers will compact under the action of the waves. In other words: the waves will cause a settlement of the armour layer, and this settlement will be greater the steeper the slope of the armour layer. What happens when the compacted armour layer is settling? Obviously, this causes a change in the contact forces between individual elements both dynamically during the compaction and statically when the armour layer units are at rest again.

Some of the contact forces will become smaller, and some will increase, but the average tendency will be towards increase.

More of the weight of the armour layer will be carried by the armour layer proper.

So we have a situation where the contact forces, and therefore the moments in the units, will increase under the impact of the waves. Considering that we now know that the strength of the concrete armour units is a very critical parameter, it is no wonder that armour units break, not only in the surface layer due to rocking, but also in the surface layer and in the bottom layer due to compaction/settlements.

For large rubble-mound breakwaters, where the concrete units become relatively weaker, breakage of elements in the surface layer due to rocking and due to compaction (settlements) occurs in both the surface layer and the bottom layer.

The compaction/settlement effect is probably the more important of the two, because there seems to be no other way to explain why up to 80% of the tetrapods in Port d'Arzew El Djedid breakwater bottom layer were broken in certain sections which were not subject to collapse when surveyed after the failure.

Example:
The large breakwater in Port d'Arzew El Djedid in Algeria failed during a severe storm in December 1980 (ref. 1). The project profile appears in Fig. 1 and two typical profiles of the breakwater surveyed after the storm are seen in Figs 2 and 3. Note that the armour

Breakwaters—design and construction. Thomas Telford Ltd, London, 1984

173

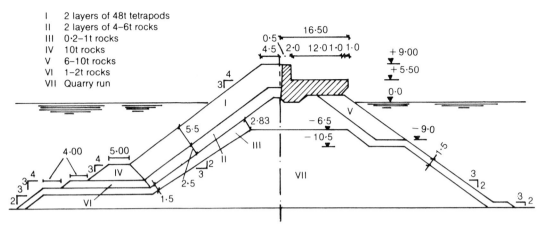

Fig. 1. Cross-section of the main breakwater in Port d'Arzew El Djedid

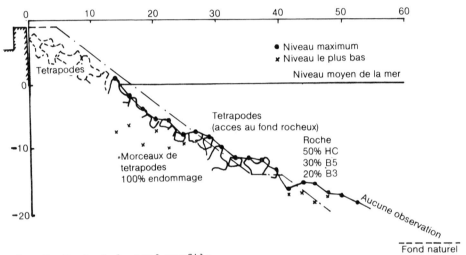

Fig. 2. Typical damaged profile

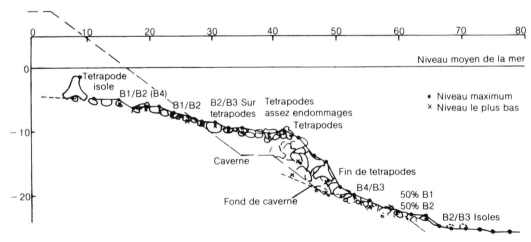

Fig. 3. Typical damaged profile in sections with complete failure/collapse

layer consists of 20 m^3 (48t) tetrapods on a slope of 1:1.33. Fig. 2 is typical for sections of the breakwater where heavy damage occurred and where a substantial percentage of the tetrapods are broken in the armour layer, while Fig. 3 is typical for sections where a complete failure of the armour layer occurred and the armour layer has slid down. This profile appears now with a gentle slope of about 1:5 down to level -10 m, below which the material (tetrapods, pieces of broken tetrapods and quarry stones from the filter layers and the core) stands almost at the angle of repose. Fig. 2 shows that the breakage of tetrapods mainly occurred from SWL down to a level of about -10 m. It is further seen that the whole armour layer has settled from its initial profile. The settlement (due to compaction) is in the order of 2-4 m at the crest and upper part of the armour layer.

According to a diver survey of the interior of the armour layer up to 80% of the tetrapods below SWL were broken in certain sections. Flume model tests with irregular waves on the project profile showed settlement/compaction of the armour layer as a whole. The settlements were on the average in the order of 2.0 (1-4 m) vertically, when measured at the wave wall. The settlements began for significant wave heights, H_s, in the range of 2.5-5.0 m in repeated test runs, which is far below the wave height for which tetrapods were displaced from the armour layer (displacement of about 2% of the units occurred for H_s = 6.5 m).

From observations in the model, it appeared that the settlements/compaction occurred during the run-up/draw-down process of a few individual waves. The whole armour layer above level -10 m approximately were simultaneously moving, causing small adjustments of the positions of each individual unit, which resulted in compaction/settling of the upper 75% of the armour layer. It is very important to notice that most often the compaction occurred without any individual tetrapods being displaced from the armour layer.

It should further be noted that model testing with reproduction of hindcasted storm conditions did not show similarity of model and prototype damage, the reason being (in my opinion) the fragility of the tetrapods in reality.

J. J. VAN DYK (*University of Technology, Delft*)
With reference to the remarks made by Mr Sørensen regarding breakage of tetrapods in the sub-layer, I suggest that tetrapods in the surface layer, may be lifted free from the sub-layer, allowing tetrapods in the sub-layer to rock and eventually break.

H. F. BURCHARTH (*Aalborg University, Denmark*)
During the discussion Mr Sørensen showed some underwater views of the damaged Port d'Arzew breakwater. It was reported that in some sections nearly all of the 48 t tetrapods were broken, but still in place. In my opinion the mechanical failure of the units most probably was caused by a combination of the following effects:

(a) Thermal stresses. 48 t tetrapods are brittle unless special measures have been taken to prevent large temperature differences during curing.

(b) Static loads due to gravity, settlement and arching. Rough estimates of such stresses show that failures can occur.

(c) Dynamic loads due to partly pulsating loads from wave generated drag, partly impact loads from units hitting each other when moved by the waves. Broken pieces do not always disappear from original position because they nest in between larger pieces.

(d) Fatigue. The critical stress range is reduced drastically even for relatively few impacts/cycles, i.e. one or a few storms can cause significant reduction in strength.

W. F. BAIRD (*W. F. Baird & Associates, Ottawa*)
Prototype monitoring is essential for the verification of physical models. Hydraulic laboratories are encouraged to verify their procedures by simulating changes that are defined by prototype monitoring.

Theme paper: The way ahead

H. F. BURCHARTH, Hydraulics and Coastal Engineering Laboratory, University of Aalborg, Denmark

INTRODUCTION

1. It is true for all structures that the design and the location affect the loads. Breakwaters are extreme examples of this thesis, since even small variations in waterdepth, orientation slope etc. produce large variations in the load. As the dominant load is stochastic in nature and exhibits extreme variations, it is not surprising that we have difficulty in reaching an acceptable design procedure. However, being a little provocative I would like to say that some of the recent failures of major breakwaters are due to gross errors, which cannot be explained away by a claim of insufficient basic knowledge. I would rather say that the available knowledge did not reach the designers on time. One way ahead is, therefore, and always has been, to ensure good communication of up-to-date knowledge.

2. In the design process we have six important areas to consider: The first is the function of the structure, which has hopefully been specified by the client. The second, the natural boundary conditions like wave climate, available materials etc. The third, the physical behaviour of the structural elements. The fourth, the constructability. The fifth, the maintenance throughout the life of the structure, and the sixth, the assessment of the structural reliability.

Function
Natural boundary conditions
Physical behaviour of the structural elements
Construction
Maintenance
Reliability

Areas to consider in the design process

I would like to give you my view on some of the ways ahead and relate them to these areas.

DESIGN ASPECTS

INTEGRATED DESIGN

3. A breakwater is never an individual structure which is designed for its own sake, it is always a part of a harbour or a coastal protection scheme. The function can, therefore, be set only by integrated design of breakwater, access channels, moorings, loading systems, calling frequency, harbour operation, storm warning systems etc. An up-to-date design involves extensive use of both mathematical and physical models, which are seldom to the hand of consulting engineering companies or the harbour authorities. Therefore, before the functions of the breakwater are specified, it is important to communicate with capable hydraulic laborties or institutes, and this should be done right at the beginning of the planning stage. We still see examples where money is wasted due to lack of, what I would call, an integrated design process in which the client, the consultant, and the specialized laboratories and institutes work parallel and hand in hand. The development in stormwarning models and mooring and cargo handling systems will possibly lead to further reduction of the necessary capabilities of the breakwater, which again means shorter and lower breakwaters, maybe no breakwaters at all in some places.

SIMPLICITY IN DESIGN

4. As long as reliable data on wave climate are not available, and as long as the basic principles of rubble mound structures are not well understood, simplicity in design is essential. Generally this means the separation of functions.

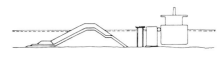

SEPARATION OF FUNCTIONS

Breakwaters—design and construction. Thomas Telford Ltd, London, 1984

177

For example it is a great advantage if the function of the breakwater is restricted to reduce transmission of wave energy. Every time we add other appendages to it, like access roads, pipeline galleries, storage areas etc., we introduce structural elements which remove it from the initially simple concept of a rubble mound. The basic rubble mound structure, in principle, is easy to construct and not sensitive to catastrophic damage if the design loads are exceeded. Wave walls are especially harmful obstacles if design waves are exceeded, because they reflect the run-up and cause increase of downrush and erosion of armour.

FAILURE MODES
5. Failures are of many types. We should be aware of the fact that we ourselves, to a great extent, can decide the mode of failure. This has been the practice for many years in earthquake engineering.

The most ductile failure is that of a traditional rubble mound with moderate sloping armour of non-interlocking blocks. The most brittle failure is associated with steep slopes of interlocking armour in front of a wave wall. The breakwater should preferably be designed for a low rate of damage, i.e. a small angle α in the figure.

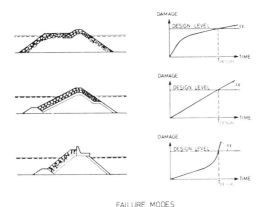

FAILURE MODES

It is essential that we design the breakwater so that the mode of failure corresponds to the quality of information constituting the boundary conditions and to the risk acceptable to the client.

WAVE CLIMATE

GROUPING
6. The more brittle the mode of failure of the breakwater, the more details of the wave climate we require.

Since oncoming waves are always interacting with the downrush from the preceeding wave, the actual succession of waves as produced in nature is important. For steep slopes with complex armour units and maybe a superstructure with a wave wall, we know that one single wave can initiate a fast progressive failure or even cause a failure by fluidising the armour layer, resulting in a slide of the armour as a whole.

7. High confidence estimates of the safety for such structures can only be achieved if we know the probability of occurrence of the wave patterns that are dangerous for the specific type of sea bed - breakwater profile.

8. One type of wave pattern, namely runs of big waves or wave groups, is generally accepted as an important wave pattern, not only for slow respons systems like moored ships, but also for rigid structures like breakwaters. However, there are other dangerous wave patterns, for example what I call a jump.

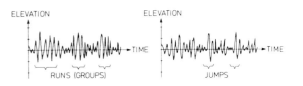

DANGEROUS WAVE PATTERNS

9. A jump is seen on the wave amplitude time series as a small wave followed by a big one. The small wave will stabilize the water table at a level close to MSL, which means that the large oncoming wave is not tripped by a downrush and therefore breaks higher up on the slope and creates higher run-up and overflow velocities. The run-down is also very deep and this causes big pressure gradients, which are destabilising.

10. At the Conference on Coastal and Port Engineering in Developing Countries the delegates had the opportunity to study the long swell on the west coast of Sri Lanka. The beach profile in some places was as steep as 1 in 5, so breaking took place virtually on the beach. Conditions were close to those for a flat rubber mound breakwater. Two phenomena made a very big impression on me. One was the tremendous drag-forces from a run-up, reaching 4 m/sec., which makes one wonder how breakwaters, where the process of erosion-accreation is not acceptable, can be stable. The other was that jumps caused much larger run-up that runs. This observation is in accordance with the run-up tests and armour stability tests I made at the Hydraulics Research Station, Wallingford, in 1977 (ref. 1).

11. To me there is no doubt that, dealing with rubble mound breakwaters with brittle failure modes, we must analyse the wave records also for wave jumps and other dangerous patterns that might be identified, and generate the laboratory waves accordingly. At University of Florida at Gainsville they are, at the moment, trying to identify such wave patterns.

12. It might be that the statistics of runs and jumps are complimentary in that, if a record contains many runs, then it contains few jumps and visa versa. An analysis of natural records from two storms and an analysis of corresponding random phase laboratory waves, which I made some years ago at the Hydraulics Research Sta-

tion, Wallingford showed this tendency, which is understandable (ref. 2). If there is such a general correlation, we probably only need to analyse the wave records for one wave pattern. This, however, still needs to be verified.

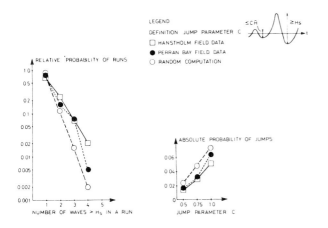

Statistics of runs and jumps (ref. 2)

13. There is some evidence that, if random phases are applied to component waves of sea, we obtain the correct statistics of wave patterns when generating the laboratory waves in accordance with the short wave spectrum (Burcharth ref. 2, Rye ref. 3, Sand ref. 4, Lundgren ref. 5). So if the hypothesis holds that groups and runs are complementary and runs are reproduced statistically correctly from the short wave spectrum with random phases, then we only need to analyse the natural wave records for the short wave spectrum and generate the laboratory waves accordingly.

14. Not all laboratories are patient enough to wait for the dangerous wave patterns in waves generated in accordance with this system. Instead they produce a deterministic amount of groupiness and hope to be on the safe side. Not all researchers believe in random phases, because natural wave records, in some cases, show more pronounced grouping than expected from random phase generation. This might be due to the limited number of waves in the available records. Twenty minutes are too short to obtain a good estimate on wave group statistic and long wave spectra.

15. Also the pronounced grouping in some records and thereby the big scatter in wave group statistics might be due to the fact that, in the analysis of wave groups, generally no distinction has been made between sea and swell. A con-

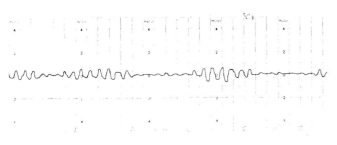

SWELL RECORD, PERRANPORTH, ATLANTIC COAST OF U K

siderable amount of swell will change the group statistics, because swell is characterized by very long runs.

From a physical point of view, it is also important to distinguish between sea and swell because, generally, they have different directions.

16. At present, a study of the correlation between runs, jumps, and short and long wave spectra is performed at Aalborg University and at the Danish Hydraulic Institute, partly based on long records of storm waves kindly provided by the Coastal Engineering Research Center, U.S. and the Royal Netherlands Meteorological Institute.

17. I would suggest that some of the many waverider buoys in service are set to one hour of recording, which allows us to calculate the long wave spectra, and that, when analysing wave groups, one look into the history of the storm to verify if significant swell components are present. This cannot always be done by looking at the variance spectrum, since significant swell might be found on frequencies close to those for the sea.

18. By advocating this separation in the analysis I am certainly not saying that swell is not important. In fact, I believe that for any breakwater on coastlines with long fetches, for instance ocean coasts, there is a strong possibility of the worst wave climate being a sum of swell and sea, as has been found for Sines (Mynett et al., ref. 6).

THREE-DIMENSIONAL WAVES
19. Swell and sea have, in general, different directions in deep water. It is, therefore, important that the direction of energy propagation is taken into account when dealing with deep water breakwaters. Also, there is some evidence from tests at the Technical University at Lyngby, Denmark (Broberg and Thunbo, ref. 7), that short crested seas affect breakwaters differently from long crested seas, which is to be expected. Therefore, reliable arrays of wave gauges to record directional seas must be developed and brought into operation as soon as possible. It should be remembered that, for many breakwaters, overtopping is the critical factor, and that the amount of overtopping and its distribution in time and space must be greatly affected by the lengths and directions of the wave crests. There is still a great deal of research to be done in this field.

FREAK WAVES
20. For deep water breakwaters we also have the problem of freak waves, that is waves of extreme dimensions, and often with unexpected direction of propagation. Ship records, through generations, mention these often catastrophic waves, and as a sailor I have myself met such a wave in the Skagerak. Such waves are difficult to implement in the breakwater design process, partly because they are so rare that records are very few and a statistical treatment therefore impossible, partly because laboratory gen-

eration of such waves is difficult. The Norwegian Hydrodynamic Laboratories (Kjeldsen, ref. 8) and the National Research Council of Canada (Mansard et al., ref. 9) are working on this problem.

21. If we cannot obtain wave information on the characteristics and the frequency of occurence of such waves, we might then use a procedure where the designer decides on a certain deterministic freak wave sequence, and the breakwater is then designed such that serious damage, but not a complete failure of the breakwater occurs, when exposed to this sequence.

PHYSICAL PROCESSES IN WAVE STRUCTURE INTERACTIONS

REVISION OF ARMOUR STABILITY FORMULAE

22. The different parameters involved in wave structure interactions are known qualitatively, but not quantitatively. Consequently our formulae for calculating wave load on structural members are primitive and somewhat unreliable. The formulae are based on dimensional analysis of a few of the parameters involved and then fitted to model-test results by applying a coefficient. This is partly a "black box" approach where the coefficient must take care of many parameters and as a result is not a constant, but a function of several variables. The classical example is K_D in the Hudson equation, which is not a constant since it varies considerably with both the slope angle and the wave period for complex types of armour. For example on a horizontal bed, Dolosse and rocks of the same weight have equal hydraulic stability (Burcharth and Thompson, ref. 10). Since it was clearly shown, already in 1974, in a paper (Brorsen et al., ref. 11) presented at the Coastal Engineering Conference in Copenhagen that K_D cannot be taken as a block specific constant, I think it is time to revise the different manuals and codes of practise in this respect, and also to point out that in many cases the only tools we have, at present, are model tests.

PARAMETER ANALYSIS

23. We must, of course, proceed from this primitive stage by a more systematic investigation of the influences of the different parameters. This can only be done by restricting the number of variables, for example as done by Brebner (ref. 12) in his steady flow tests with Dolosse, by Alan Price (ref. 13) in his pull-out tests and by Burcharth and Thompson (ref. 10) in the pulsating water tunnel tests with beds of Dolosse and rocks. Such tests lead to an understanding of the relative importance of a few parameters at a time. There are many possibilities for such tests, and I will for research recommend more such deterministic tests instead of tests with complete breakwater slopes, where the relative influence of different parameters is difficult to verify, although it is possible in some cases as shown for example by Bruun and Johanneson (ref. 14), and Gravesen and Torben Sørensen (ref. 15).

MODEL EFFECTS

24. Since we have to rely on model tests a great deal, it is very important to consider both model effects, recording techniques and scale effects. By model effects I mean those due to deficiencies in the model, for example the use of two-dimensional waves instead of three-dimensional waves, improper reproduction of wave trains, surface roughness of armour blocks etc. Model effects can be considerable. Examples are models of shallow water breakwaters where a fairly steep bottom profile in front of the breakwater is not incorporated into the model. We know that even small variations in the bottom profile cause considerable changes in the waves attacking the structure. Where changes of profile along the breakwater are present, the model test program often has to be extended to cover a substantial part of the sea bed topography. This has to be accepted, although it will result in a considerable increase of model costs.

SCALE EFFECTS

25. Scale effects in rubble mound models are still discussed at length, for example by Vasco Costa (ref. 16), and for good reasons. There are two types of scale effects which especially need clarification. One is related to the pressure from breaking waves, the other to flow in the porous body.

26. Wave pressures on superstructures and armour units will in a Froude model suffer from scale effects, partly due to incorrect amount and distribution of entrapped air, partly due to wrong scaling of air compressability. Moreover, it is indicated by Joe Ploug, the chairman of the IAHR working group on wave generation and analysis, that wave models, although built and run to the same detailed specifications, show significant differences in various respects. It is, therefore, to be expected that the amount of entrapped and saturated air will vary significantly in models of the same prototype but produced in various laboratories. Even though the part of the wave pressure, which in the model is recorded as shock pressure, can be interpreted in accordance with the compressibility law, and even though it might be argued that the model test results are on the safe side, the fact still remains that we do not know how close we are to reality.

27. A research project with the participation of a number of laboratories should be started with the object of studying the pressures on wave walls and capping base plates. Tests in long and short crested waves as well as tests in wind wave flumes are needed. An important part

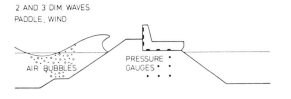

2 AND 3 DIM. WAVES
PADDLE, WIND

AIR BUBBLES

PRESSURE
GAUGES

WAVE PRESSURE SENSITIVITY ANALYSIS

of the study should be tests to verify how sensitive pressures are to the amount of saturated and entrapped air. This might be done by adding air bubbles. As the most important part of the project, prototype recordings of pressures should be performed for the comparison and calibration of the model test results.

28. The other important scale effect is the viscous effect in porous flow. We know that there are no significant viscous scale effect in flow in armour layers even in small scale models (Burcharth and Thompson, ref. 10; Mol, Ligteringen et al., ref 17). However, this flow in the armour layer is affected by the flow in the filter layers and the core, where viscous effects might be present due to much lower permeability. This is easily visualized by the change in positions of the internal water table. Generally a viscous effect will result in too high an internal water table.

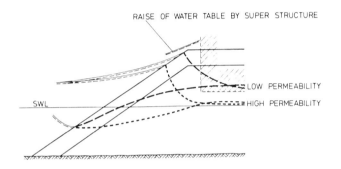

INFLUENCE OF PERMEABILITY ON INTERNAL WATER TABLE

For the armour it means larger destabilizing pressure gradients and a reduction in the valuable reservoir effect, thus giving rise to larger overflow velocities. For the superstructure too high a position of the internal water table might enlarge the wave pressures, both on the wave wall and the base, to such an extent that the results will be completely wrong.

29. A generally applicable method of preventing viscous effects is to assure that the Reynold's number for the armour layer flow exceeds a certain value (Dai and Kamel, ref. 18). However, this criterion is not satisfactory, first of all because a single value of a Reynold's number cannot, in general, represent the complicated and unsteady flow field, where, even in prototype, some part of the flow will most probably be laminar or fluctuating between turbulent and laminar. This problem has recently been evaluated by Juul Jensen and Klinting (ref. 19). Another problem is that the permeability of prototype cores is very difficult to predict since the permeability is sensitive to small variations in the grading and also to separation of material when dumped. The prototype flow field is, therefore, generally poorly known. A very promising mathematical model of the porous flow has been developed by Barends et al. (ref. 20) at the Delft Soil Mechanics Laboratories, but need to be calibrated against prototype data.

30. An important step in solving the problem is, therefore, to obtain prototype data on the pressure field inside various typical breakwaters for the calibration of both physical and mathematical models.

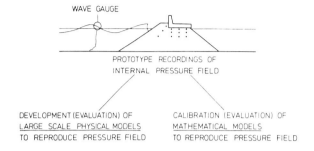

31. To get the mathematical models operational is very important, since such models are the easiest tools for estimating the internal pressure field and water table positions in new designs. Based on experience gained by prototype measurement it might be fairly easy to develop a technique for the construction of good physical large-scale models, but for the usual smaller scale models the solution might be, first to run a mathematical model for the determination of the internal pressure field and thereafter, by varying the core permeability, to calibrate the small-scale physical model to reproduce the calculated internal pressure field. When this is done, then other phenomena, like run-up, armour stability etc., can be studied, unbiased by viscous effects.

Since measurements of pressures only require fairly simple instrumentation, it should be possible to realize such prototype recordings in new breakwaters where also wave recording takes place.

FORCES ON ARMOUR UNITS

32. Although recording techniques have developed considerably in recent years, we still for armour units require a good method of recording movements, or preferably forces or stresses. For this reason we are still missing the hydrodynamic and mechanical response functions for armour units exposed to waves. These functions are the missing link in the design procedure where both the hydraulic and the mechanical stability are considered.

33. We can find the respons of individual blocks to deterministic well defined loads. For example from theoretical considerations and full scale experiments, we can obtain information on the relative variation and, to some extent, the absolute variation in strength with size of units (Burcharth, ref. 21 and 22; Silva, ref. 23).

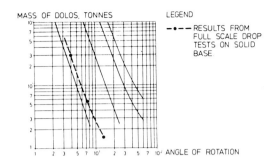

MASS OF DOLOS, TONNES

LEGEND

—•—RESULTS FROM FULL SCALE DROP TESTS ON SOLID BASE

ANGLE OF ROTATION

EXAMPLES OF ISO-STRESS GRAPHS FOR ROCKING UNREINFORCED DOLOSSE OF IDENTICAL CONCRETE WAIST RATIOS 0.30-0.36

The figure shows an example of such approach for Dolosse. The family of curves are iso-stress lines and the dotted curve is one found from droptests on a hard base. By relating to proto-type experience, we can obtain an indirect design method (Burcharth, ref. 24).

34. However, a more direct approach for the determination of loads on armour units is wanted. The detection of armour unit movements is still an indirect method, but valuable. Photo technique in models can, to some extent, be used to detect the movements. However, it fails in the splash zone, which, unfortunately often, is very critical and therefore most important. Moreover, the rocking stage, where no displacements take place, cannot be detected by single frame technique, and this stage is very important for some un-reinforced slender concrete units.

35. The use of accelerometers or stain gauges in model armour represents a more direct method and is a big step forward, and has already been implemented as a standard procedure, for example at the Delft and De Voorst Laboratories. The instrumentation of armour units is expensive and restricted to fairly large units. As a result only a small number of units in rather few models have been instrumented so far. But to obtain reasonable confidence levels we need a whole pack in each model, say minimum 50 units, instrumented to give simultaneous readings. To minimize scale effects, this should be done in some of the giant flumes. Also, it is very important that we include a number of instrumented armour units in new breakwaters where wave gauges are installed.

36. Another way of dealing with forces on armour units in a pack is to scale the material characteristics of the concrete in accordance with the Froudian Law of similitude and record the damage. Despite the very difficult task of scaling both tensile strength, compressive strength, modulus of elasticity and fracture toughness, Gerry Timco (refs. 25 and 26) has been very successful in his work with model concrete for small scale models. It is, of course, much easier when working at larger scale, for example length scale 1:10, which is possible in the giant flumes, and model concrete has been used in the Delta flume at De Voorst in the large scale models, for example of the redesigned Sines breakwaters. However, the modelling of the concrete can still be improved.

37. For the testing of specific breakwater designs the application of correctly scaled model concrete is the best procedure. From a research point of view, a more promising approach might be to run parallel tests with armour units made of relatively weaker materials, but with different characteristics, so that mechanical failure of armour units takes place at various stages, also before hydraulic instability is expected. In this way, we can get information on the distribution in space and time of static and pulsating loads in a pack.

38. A different approach is, of course, to use a mathematical model. Various parts of such a model can be determined and checked in physical models. For example the hydrodynamic response function of a unit in a pack might be determined by fixing the unit to the end of a flexual beam mounted with strain gauges and arranged in the pack such that the surrounding units (which must be glued to prevent movements) are not touched. The response, in terms of total force, to different flow characteristics could then be used as input in a mathematical model for the estimate of the distribution of forces on the units. Such a mathematical model must be able to construct a geometrically correct pack of armour units, to calculate the flow field from complicated boundary condition, to calculate the flow force on individual units, and to calculate the magnitude and direction of the contact forces between the units. Also, it must take into account the reallocation of contact forces when local crushing and interblock sliding take place. A promising model, able to handle geometrically simple shaped elements, is reported by Austin and Schluter (ref. 27), and no doubt models which can handle also geometrically complicated shaped armour will soon be available, but to my opinion not applicable as design tool if not calibrated against physical data. Especially information on crushing characteristics in contact points is essential.

39. All four above-mentioned approaches for studying the forces on armour units will be necessary if, in our design process, we want to be able to also design armour units as we are designing other structural elements.

INSTRUMENTATION OF ARMOUR UNITS IN LARGE SCALE MODELS

INSTRUMENTATION OF PROTOTYPE ARMOUR UNITS

STUDY OF "WEAK-MATERIAL" ARMOUR UNIT FAILURES IN MODELS

MATHEMATICAL MODELS

WAYS OF DETERMINING LOADS ON INDIVIDUAL ARMOUR UNITS IN A PACK

The results are needed, not only as an important brick in a consistent design method, but also because we want to be able to test breakwaters in scale models without the expensive scaling of the armour material. Because of the costs, such a project is an obvious subject for international sponsorship. A working group under PIANC should be considered.

ARMOUR UNITS

DESIGN ASPECTS

40. The development of the different types of concrete armour units was based mainly on practical experience. Only in a few cases, a more systematical design process based on both hydraulic and structural consideration has been applied. New types of units will certainly be developed parallel to progress in the understanding of hydraulic and structural behaviour of armour. At present the different types of concrete armour units might be put into the following three categories: Bulky types randomly placed (cubes, Antifer blocks), complex slender types randomly placed (Tetrapods, Dolosse), and hollowed blocks placed in patterns (Cob, Shed, Diode).

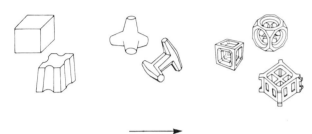

INCREASING ENERGY DISSIPATION TO MASS RATIO

CATEGORIES OF CONCRETE ARMOUR UNITS

41. We know that, from a hydraulic point of view, a high permeability is most advantageous (Burcharth and Thompson, ref. 10). The introduction of the ROBLOC sublayer unit by P.R.C. Harris (Groeneveld et al., ref. 28) to ensure random placement and thereby large permeability of cube armour follows this line. This means as little concrete as possible. We also know that randomly placed units involve impact forces due to movements, and also large variations in contact forces.

42. Hollowed cube types have a high permeability and exhibit practically no movements when correctly placed like a pavement, and the contact forces are uniformly distributed. This together with a structurally better shape makes the hollow types the most efficient in terms of the optimum use of concrete for dissipation of wave energy (Barber et al., ref. 29; Wilkinson et al., ref. 30). This type of unit is widely used for sea wall revetments, where construction at low tide can be done in the dry, because it demands an even underlayer, accurately placement and good toe support. However, as the technique for the construction and the control of underwater works improve, I believe we shall see the hollowed cube types of units being used also in breakwaters.

43. The slender, complex types of units, which at the moment seem to be less popular because of the recent failures, will not disappear from the scene, but the use of such units will probably be restricted to breakwaters where the wave climate can be well predicted, as for example in shallow water situations. Because of the units' good hydraulic stability relative to the mass, such units will again be used for deep water structures when a proper design method, which takes into account also the mechanical properties of the units, is developed. However, there is no doubt that the static and dynamic strength of the larger size slender complex types of units must be increased to allow for some movements, because such units cannot compete with more bulky types if non-movement criteria are adopted, and also because movements are unavoidable in a randomly placed pack.

44. A properly designed armour unit must have a homogeneous response to the various loads. For an analytical analysis of stresses we need not only the hydrodynamic transfer functions, as mentioned above, but also the material response. Elastic and plastic theories as well as fracture mechanics have proved to be inadequate in describing the mechanical response of concrete, for example the fatigue. Much more promising is the Continuous Damage Theory by Dusan Krajcinovic (refs. 31, 32, 33) and others. This is not very surprising, since the theory is based on the introduction of an additional kinematic variable, characterizing the density and the distribution of the microcracks and their effect on the material parameters and the stress redistribution.

CONCRETE

45. On the material side we are looking for a less brittle concrete with good long term durability. We know that the use of super-plasticizers, and thereby a low water cement ratio, and the use of puzzolan cement is beneficial in this respect. Full scale static and dynamic tests by Polytechna Harris, Milan, with 30 t Dolosse in Gioia Tauro clearly confirmed this.

46. When dealing with brittle materials like concrete it is essential to prevent tensile stresses in the surface zones to develop during production. Even small tensile stresses in the surface regions will seriously affect the strength of the member. On the other hand, if we can assure the reverse, namely compressive stresses in the surface zone, then a considerable increase in structural strength might be obtained. Such a technique is known from production of, for instance, glass and might be possible also for concrete. For example, by ensuring more pronounced hydratization and thereby expansion in the surface regions than in the centre regions.

47. Progress will be made if we can get the concrete researchers interested in finding ways of improving especially the tensile strength and the fracture toughness. For many years these properties have not attracted much interest from researchers because structural elements exposed to tensile stresses are normally reinforced.

FATIGUE

48. Armour units are exposed to repeated loads, typically several million cycles at various

stress levels during the structural lifetime. The loads are both pulsating and impacting. Recent research by Tepfers (ref. 34), Fagerlund et al. (ref. 35), Zielinsky et al. (refs. 36, 37), Tait (ref. 42), and Burcharth shows a significant fatigue effect in concrete, which clearly must be incorporated in the design procedure.

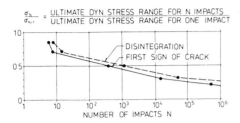

$$\frac{\sigma_N}{\sigma_{N,1}} = \frac{\text{ULTIMATE DYN STRESS RANGE FOR N IMPACTS}}{\text{ULTIMATE DYN STRESS RANGE FOR ONE IMPACT}}$$

FATIGUE IMPACT LOADED FLYASH CONCRETE DOLOSSE FLEXURAL STRESS (PRELIMINARY RESULTS, BURCHARTH, 1983)

For example the allowable flexural stress range is reduced to 40% after approximately 10,000 impacts or cycles only. Improvement of the ductility and fracture toughness, as mentioned previously, will increase the fatigue life also, but much more research in this field is needed.

REINFORCEMENT

49. Reinforcing the concrete is, of course, an obvious way of improving the strength properties. Both conventional steel bar reinforcement and fibre reinforcement are used. Results from full scale static tests and dynamic drop tests with Dolosse in the range 1.5 t - 30 t indicate that conventional reinforcement is superior to steel fibres of equal quantity. By using approximately 130 kg steel per m^3 concrete, spalling and not cracking seems to be the limiting factor (Burcharth, ref. 21, Polytechna Harris, Milano 1983). Fibre is beneficial in very slender and complicated structured members such as hollowed cube types of units. Chopped polypropylene fibres are, for example, used successfully in the SHED unit by Shephard Hill Ltd. Dolosse and Tetrapods are, in this respect, not slender but relatively bulky and stiff elements.

The cost of bar reinforcement will be reduced as more effective ways of using the steel and easier ways of placing the reinforcements are developed. Ongoing research in this field are full scale tests and finite element calculations by Mike Uzumeri at University of Toronto.

CORROSION

50. Corrosion has prevented many coastal engineers from using steel reinforcement. Research on corrosion is intense and promising. Results obtained so far at the Danish Corrosion Centre (F. Grønvold) show that the use of fly ash reduces corrosion, and high densified concrete with a substantial content of silica dust nearly eliminates the risk of harmful corrosion in bars of the sizes used in large armour units. The influence of crack width on corrosion is still not fully understood. The use of reinforced concrete

in offshore structures will ensure further research and thereby improvements applicable also in breakwater concrete technology.

THERMAL STRESSES

51. Thermall stresses represent a serious problem for all armour units of large volume. Roughly, it can be said that micro cracking will occur if temperature differences during curing exceed 20 °C. This compares to dimensions exceeding approximately 1m x 1m x 1m, when conventional concrete and casting procedure are used. The problem, therefore, exists also for the larger complex types of units. The figure shows a calculation by the Danish Concrete Institute BKI of the extent of microcracking and stresses in a 70 t cube produced by conventional concrete technology. No wonder that it is found that such cubes often are very brittle.

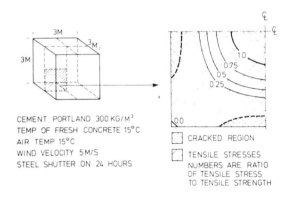

CEMENT PORTLAND 300 KG/M³
TEMP OF FRESH CONCRETE 15°C
AIR TEMP 15°C
WIND VELOCITY 5 M/S
STEEL SHUTTER ON 24 HOURS

CRACKED REGION

TENSILE STRESSES NUMBERS ARE RATIO OF TENSILE STRESS TO TENSILE STRENGTH

THERMALL STRESSES IN A 70 T CUBE 100 HOURS AFTER CASTING (BKI-INSTITUTTET COPENHAGEN AND BURCHARTH 1982)

Measures to prevent thermall stresses are well known, but they all involve drawbacks. The use of low heat cement or retarder slows down production, the use of less cement reduces the surface resistance and the long term durability, cooling of aggregates and water are expensive and impossible in some places, and the use of insulation during the curing complicates the production. This is illustrated in the figure, which shows an example of a diagram by BKI-Instituttet and Burcharth for determining the number of days where insulation must be kept on a

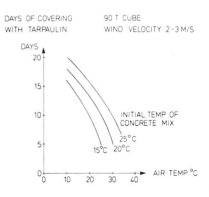

DAYS OF COVERING WITH TARPAULIN

90 T CUBE WIND VELOCITY 2-3 M/S

INITIAL TEMP OF CONCRETE MIX

EXAMPLE OF DIAGRAM TO DETERMINE DURATION OF INSULATION DURING CURING (BKI-INSTITUTTET COPENHAGEN AND BURCHARTH, 1982)

90 t cube of conventional concrete to prevent
thermall cracking. It is seen that approximately
15 days are necessary, which again demands 300-
1000 insulation sets, depending on the size of
the job.

53. For some units it is easy to solve the
thermall stress problem by adjusting the shape.
The figure shows how this can be done for a big
antifer cube, simply by making a hole in the
middle. In addition, such a modification will
increase the hydraulic performance of the armour.

ANTIFER TYPE BLOCK WITH HOLE
TO REDUCE THERMALL STRESSES

COMPOSITE TYPE OF ARMOUR UNIT

54. Many of the problems related to armour
units might be solved by a composite type of
unit, consisting of a thin shell of very strong,
ductile fibre reinforced cement and filled with
mass concrete with low cement content or low
heat cement.

PROPOSAL FOR COMPOSITE ARMOUR UNIT
(PRINCIPLE SHOWN FOR DOLOS-SHAPED UNIT)

The concrete can be of an inferior quality than
traditionally specified. This solves many problems in areas where good qualities of aggregates,
cement and water are not available.

A good surface resistance and long term durability will be ensured by the high quality shell.
The production will give few limitations to the
shape because the fibre reinforced mortar shell
can be manufactured by a spraying technique.

The edges of the two parts of the shell can be
designed with shear keys. The feasibility of
such a composite unit is not known. Besides the
economy, we must study, both theoretically and
experimentally, the static strength and the
fatigue life before anything can be said.

SURFACE ROUGHNESS

55. To obtain high hydraulic stability, the
shape of the unit should ensure a very high permeability of the armour layer. For this reason

also a very high surface roughness is desirable.
This can be obtained by shaping the outer shell
with fairly big roughness elements. This will
also increase the stiffness of the shell so that
pressure from the wet concrete does not cause
too big deformations.

ROUGHNESS ELEMENTS ON ARMOUR UNITS

The positive effect of surface roughness is evident from a breakwater I saw on the Maldives and
which was constructed of coral blocks. The mass
of the coral blocks is approximately 20 kilos,
the density only 1.1 t/m^3 and the front and back
slopes nearly vertical, and still this breakwater resists long waves of more than 2 m's
height. The roughness made it impossible for me
to pull a block out from the pack, which I also
tried in vain from a groin, constructed in the
same way.

MONITORING OF UNDERWATER WORKS DURING CONSTRUCTION AND STRUCTURE LIFETIME

56. Many breakwater designs involve such construction difficulties that a not-as-designed
structure is the inevitable result. Fracture of
fragile armour units during placement and not-
as-designed distribution of materials in underwater sections are the most serious problems.
Although we can improve the strength of armour
units and put more attention to constructability
in the design, the two mentioned problems will
still remain to some extent.

57. Fracture of armour units by underwater
placement is due to high impact velocities. The
velocity must therefore be recorded, for example
by placing accelerometers on the sling (grab),
and transmitted to the crane operator.

58. To secure a uniform distribution and placing
density of underwater armour for some types of
units, it is necessary to use a technique of
visual observation. In large water depths it is
not sufficient to place units to a co-ordinated
grid, because even a moderate swell might displace the unit a couple of metres, which cannot
be registered by the operator. Moreover, a satisfactory placement also requires a certain
orientation of the unit, when placed to fit in
the pack. The orientation and the rotation of
the unit should therefore be visualized and controlled by the operator. Underwater low light
colour video cameras mounted on a direction
stabilized frame above the sling (grab) might
solve the problem.

59. The inspection of the breakwater profile
and the control of the coverage density (gabs

in the pack) are very important but difficult, especially in the case of big complex types of blocks. Side scan sonar has been tried with moderate success (Patterson and Pope, ref. 38). However, experts say that already available and not too expensive technique makes it possible to obtain detailed pictures, for example by presenting densely spaced profiles produced from position corrected signals from a scanning sonar mounted on a boat. The figure shows a representation for a sea bed.

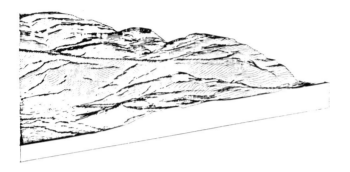

EXAMPLE OF 3D REPRESENTATION FROM MONITORED PROFILES (EIVA LTD, DENMARK)

Devices for the monitoring of underwater works for rubble mound breakwaters can be developed further. Since the costs involved will be minor compared to the construction costs, it is obvious that the designer should stimulate such a development, simply by specifying strict procedures for the control of underwater works in the tender documents. Without development in this field it is also impossible to follow the needed programmes for underwater maintenance throughout the lifetime of the structure.

PROBABILISTIC DESIGN

61. Very often the argument is heard that a probabilistic design procedure is of little value as long as the understanding of the physics is poor. It is true that such a design process never gives you figures in which to place high confidence as long as we cannot describe the physical processes. However it is worth while to recall that the less we know about the physics, the more important it is to try to assess the reliability. The only method is the probabilistic approach, which gives you information on the risk of failure with due consideration to the uncertainty or scatter of the various parameters involved. It is no excuse not to use the method because we do not know the probability density functions of the individual parameters (or of the traditionally used black box parameters). As engineers we must estimate the distributions, just as we estimate partial coefficients or safety factors. To-day's knowledge makes it difficult to estimate the probability functions for some parameters. The advantage of the method in this respect is that you can easily vary the probability distributions and see the influence on the reliability of the structure. Also, such a sensitivity analysis is essential in providing a systematic base which the designer can use for

the specification of the accuracy or confidence limits, for example for the model test results and the construction tolerances. (Kooman, ref. 39). Another advantage is that long term effects, such as for example fatigue, can be included in a meaningful way (Nielsen and Burcharth, ref. 40). It is, in fact, my opinion that we should use a level III method straight away with the estimation of unknown distribution functions. It would of course be most valuable if the various laboratories could produce and report estimates of parameter distributions.

62. The probabilistic method cannot, in an operational way, take into account the so-called gross errors, which in fact very often are responsible for unexpected failures. Gross errors are, for example, if the designer forgets or does not know about the mechanical strength problem for slender unreinforced units, or if the designer produces design sea state on wave data belonging to statistically different populations, or if the contractor puts the reinforcement bars in the wrong part of a structural member. We cannot take such gross errors into account. The evaluation of gross errors in terms of Fuzzy sets might be an operational way for the implementation in the design process (Ditlevsen, ref. 41). Another aspect of the gross error problem might be illustrated by the formula by Robin S. Colquhoun for the real probability of failure,

$$\text{REAL PROB OF FAILURE} \quad P_{F,\,real} = 1 - \exp\left[-P_f^{\,P_E}\right]$$

P_f is the engineer's estimate of prob of failure

P_E is the prob that the engineer knows what he talks about.

(By Robin S Colquhoun)

ACKNOWLEDGEMENT

The many inspiring discussions with researchers, designers, and constructors working in the coastal engineering field are gratefully acknowledged. A special thank to Mr. W.A. Price for commenting on a draft of this paper.

REFERENCES

1. BURCHARTH H.F. The effect of wave grouping on on-shore structures. Coastal Engineering, 2 (1979) 189-199.
2. BURCHARTH H.F. A comparison of nature waves and model waves with special reference to wave grouping. Coastal Engineering, 4 (1981) 303-318. Proc. Int. Conf. on Coastal Engineering, Sydney.
3. RYE H. Ocean wave groups. Ph.D. Thesis. Div. of Marine Hydrodynamics, The Norwegian Institute of Technology, Trondheim. Report VR-82-18. 1981.
4. SAND S.E. Wave grouping described by bounded long waves. Ocean Engineering, Vol. 9, No 6, 1982.

5. LUNDGREN H. Trends in coastal and port engineering research. Keynote address. Int. Conf. on Coastal and Port Engineering in Developing Countries, Colombo, 1983.

6. MYNETT A.E., de VOOGT W.J.P., SCHMELTZ E.J. West Breakwater - Sines wave climatology. Proc. Coastal Structures' 83, Washington 1983.

7. BROBERG P.C., THUNBO CHRISTENSEN F. Stability of rubble mound breakwaters in 2-D and 3-D waves (in Danish). M.Sc.-thesis. Inst. Hydrodyn. and Hydraulic Engrg. Tech. University, Lyngby, Denmark. (1983) (P. Tryde and S.E. Sand supervisers).

8. KJELDSEN S.P. 2- and 3-dimensional deterministic waves in a sea. 18th Int. Conf. on Coastal Engineering, Cape Town. Abstracts, paper No 165 1982.

9. MANSARD E.P.D., FUNKE E.R., BARTHEL V. A new approach to transient wave generation. 18th Int. Conf. on Coastal Engineering, Cape Town. Abstracts paper No 167, 1982.

10. BURCHARTH H.F., THOMPSON A.C. Stability of armour units in oscillatory flow. Proc. Coastal Structures' 83. Washington 1983.

11. BRORSEN M., BURCHARTH H.F., LARSEN T. Stability of dolos slopes. 14th International Conference on Coastal Engineering. Copenhagen 1974.

12. BREBNER A. Performance of dolosse blocks in an open channel situation. Proc. 16th International Conference on Coastal Engineering. Hamburg 1978.

13. PRICE W.A. Static stability of rubble mound breakwaters. Dock & Harbour Authority. Vo LX no 702, 1979.

14. BRUUN P., JOHANNESSON P. Parameters affecting the stability of rubble mounds. Proc. ASCE Journal Waterways, Harbours and Coastal Engineering Division. Vol. 102 no WW2, 1976.

15. GRAVESEN H., SØRENSEN T. Stability of rubble mound breakwaters PIANC. 24th Int. Navigation Congress. Leningrad 1977.

16. VASCO COSTA F. Forces associated to different fluid properties as affected by scaling. Coastal Engineering, 5 (1981) 371-377.

17. MOL A, LIGTERINGEN H, GROENEVELD R.L., PITA C.R.A.M. West breakwater - Sines. Study of armour stability. Proc. Coastal Structures' 83. Washington, 1983.

18. DAI Y.B., KAMEL A.M. Scale effect tests for rubble-mound breakwaters. Research report H-69-2, U.S. Army Corps of Engineer. Waterways Experiment Station, Vicksburg 1969.

19. JUUL JENSEN O., KLINTING P. Evaluation of scale effects in hydraulic models by analysis of laminar and turbulent flow. Submitted for publication in Coastal Engineering, Elsevier, Amsterdam 1983.

20. BARENDS F.B.I., Van der KOGEL H., VIJTTEWAAL F.E., HAGENAAR J. West breakwater - Sines. Dynamic-geotechnical stability og breakwaters. Proc. Coastal Structures' 83. Washington 1983.

21. BURCHARTH H.F. Full-scale dynamic testing of Dolosse to destruction. Coastal Engineering, 4 (1981) 229-251.

22. BURCHARTH H.F. Comments on paper by G.W. Timco titled "On the structural integrity of Dolos under dynamic loading conditions". Coastal Engineering, 7 (1983) 79-101.

23. SILVA M.G.A. On the mechanical strength of cubic armour blocks. Proc. Coastal Structures' 83, Washington, 1983.

24. BURCHARTH H.F. A design method for impact loaded slender armour units. ASCE International Convention New York 1981. Laboratoriet for Hydraulik og Havnebygning, Bulletin no 18, University of Aalborg, Denmark. 1981.

25. TIMCO G.W. The development, properties and production of strength-reduced model armour units. Laboratory Technical Report, Nov. 1981. Hydraulics Laboratory Ottawa, National Research Council, Canada. 1981.

26. TIMCO G., MANSARD E.P.D. Improvements in modelling rubble-mound breakwaters. 18th Int. Conf. on Coastal Engineering, Cape Town. Abstracts paper no 145. 1982.

27. AUSTIN D.I., SCHLUTER R.S. A numerical model of wave-breakdown interactions. 18th Int. Conf. on Coastal Engineering, Cape Town. Abstracts paper no 147. 1982.

28. GROENEVELD R.L., MOL A., ZWETSLOOT R.A.I. West breakwater - Sines. New aspects of armour units. Proc. Coastal Structures' 83. Washington, 1983.

29. BARBER P.C., LLOYD T.C., STANGER L. Leasowe Revetment Reconstruction, Stage III. Report, Director of Engineering Services. Metropolitan Borough of Wirral. 1981.

30. WILKINSON A.R., ALLSOP N.W.H. Hollow block breakwater armour units. Proc. Coastal Structures'83. Washington, 1983.

31. KRAJCINOVIC D. Distributed damage theory of beams in pure bending. J. of Appl. Mech., Vol. 46, September 1979.

32. KRAJCINOVIC D., FONSECA G.U. The continuous damage theory of brittle materials, Part I: General Theory. J. of Appl. Mech., Vol. 48, December 1981.

33. KRAJCINOVIC D., FONSECA G.U. The continuous damage theory of brittle materials, Part II: Uniaxial and plane response modes. J. of Appl. Mech., Vol. 48, 1981.

34. TEPFERS R. Tensile fatigue strength of plain concrete. ACI-Journal August 1979.

35. FAGERLUND G., LARSSON B. Betongs slaghallfasthed (in Swedish). Report of Cement och Betonginstitutet, Stockholm 1979.

36. ZIELINSKI A.J., REINHARDT H.W., KÖRMELING H.A. Experiments on concrete under repeated uniaxial impact tensile loading. RILEM Matériaux et Constructions, Vol. 14, No 81, 1981.

37. ZIELINSKI A.J., REINHARDT H.W. Stress-strain behaviour of concrete and mortar at high rates of tensile loading. Cement & Concrete Research 1982.

38. PATTERSON D.R., POPE J. Coastal applications of side scan sonar. Proc. Coastal Structures' 83. Washington 1983.

39. KOOMAN D. Riskanalysis and design criteria in breakwater design. Proc. Int. seminar on criteria for design and construction of breakwaters and coastal structures. 1980.

40. NIELSEN S.P.K., BURCHARTH H.F. Stochastic design of rubble mound breakwaters. Proc. 11th IFIP Conf. on System Modelling and Optimization, Copenhagen, 1983.

41. DITLEVSEN O. Formal and real structural safety Influence of gross errors. Structural reliability (ed. P. Thoft-Christensen), Aalborg University Centre, Denmark, 1980.

42. TAIT R.B. Private correspondance (1980).